Interactive Visualization and Plotting with Julia

Create impressive data visualizations through Julia
packages such as Plots, Makie, Gadfly, and more

Diego Javier Zea

BIRMINGHAM—MUMBAI

Interactive Visualization and Plotting with Julia

Publishing Product Manager: Devika Battike
Senior Editor: Nazia Shaikh
Content Development Editor: Manikandan Kurup
Technical Editor: Rahul Limbachiya
Copy Editor: Safis Editing
Project Coordinator: Farheen Fathima
Proofreader: Safis Editing
Indexer: Hemangini Bari
Production Designer: Roshan Kawale
Marketing Coordinator: Priyanka Mhatre

First published: August 2022
Production reference: 1290722

Published by Packt Publishing Ltd.
Livery Place
35 Livery Street
Birmingham
B3 2PB, UK.

ISBN 978-1-80181-051-7
www.packt.com

I dedicate this book to my family, especially Elin and Gaël, who have been close to me during the journey, making it possible. And to the Julia community and their open source developers who do their best to make Julia grow.

Contributors

About the author

Diego Javier Zea is a contributor to the Julia Plots ecosystem and developer of MIToS, a Julia package for studying protein sequence and structure in the Julia language. He holds a Ph.D. in bioinformatics and has worked as a postdoctoral researcher at the Leloir Institute Foundation in Buenos Aires and Sorbonne Université and Muséum National d'Histoire Naturelle in Paris. He is now an Assistant Professor at Université Paris-Saclay, studying protein structure, interactions, and evolution.

About the reviewer

Ronan Arraes Jardim Chagas received a B.Sc. degree in control and automation engineering from the University of Brasília (UnB), Brazil, in 2008, and a Ph.D. in systems and control from the Aeronautics Institute of Technology (ITA), Brazil, in 2012. Since 2013, he has been with the National Institute for Space Research (INPE) as a Space Systems Engineer. He was the mission architect and technical manager of the attitude and orbit control subsystem of the Amazonia-1 satellite. During this project, he used the Julia language to perform numerous analyses, creating important tools such as the package SatelliteToolbox.jl. His main research interests include signal processing, space systems, attitude and orbit control, and artificial intelligence.

Table of Contents

6

Creating Statistical Plots

7

Visualizing Graphs

8

Visualizing Geographically Distributed Data

Section 3 – Mastering Plot Customization

Preface

Plotting and visualizing data is an everyday skill for researchers and data scientists that is helpful for discovering and communicating insights. Additionally, adding interactivity to our visualization tasks enhances data exploration steps. At the same time, interactive visualizations and animations allow us to share more compelling stories from data. This book will teach you how to perform such tasks using the Julia programming language and the packages in its ecosystem.

The Julia plotting ecosystem is a breath of fresh air for plotting and data visualization, a new ecosystem to explore. This book will guide you through that exploration. First, it will introduce you to the basics of Julia and the `Plots` package. This package allows you to write plotting specifications that you can later render using different backends without changing your code. For example, you can render the same plot using `Plotly` for JavaScript-based interactivity or `PGFPlotsX` for generating a LaTeX-based publication-quality output. These are only two examples; in this book, we will discuss the different backends and learn how to choose the most appropriate for your task. Then, the book will also introduce you to *Makie*. This plotting library shines because of its native interactivity support and GPU-powered three-dimensional plots. Another interesting feature of those plotting packages is their extensibility. Plots and Makie offer recipe systems to allow us to create new plot types and specify how to display data types. The Julia package ecosystem intensively uses these recipes to define the different visualizations we will explore in the book's second part, for example, to create classical statistical plots and visualize networks and geographically distributed data. Finally, in the book's third part, we will present the aspects needed to customize your plots to make them look as you wish. In that section, we will explore another key feature of Plots and Makie: their powerful layout system. Those systems allow for programmatically creating plot panels and figures in an easy way.

By the end of this book, you will be able to visualize data and create interactive plots with Julia. You will know how to make classical plots and develop new plot types that best suit your use case.

Who this book is for

This book is for researchers and data analysts looking to explore Julia's data visualization capabilities. It will also suit people wanting to acquire or improve plotting skills with the language. This book will support anyone developing static or interactive plots with Julia. You need basic computer programming knowledge to follow this book. However, the book does not require previous experience with the Julia language as it introduces Julia's syntax and concepts as needed. Therefore, this book could also help people who are new to the language and that are willing to start exploring its plotting ecosystem.

What this book covers

Chapter 1, An Introduction to Julia for Data Visualization and Analysis, introduces the minimal Julia concepts needed to create basic plots using the Plots package, in particular, heatmaps and scatter and line plots. To that end, it will introduce basic data types that serve as input for plotting functions. This chapter will cover the use of packages and how to work with reproducible project environments. It also presents different Julia developing environments. Among those, we will start working with Pluto notebooks—heavily used through this book. All in all, this chapter offers the essential knowledge needed for the rest of the book.

Chapter 2, The Julia Plotting Ecosystem, presents the Julia plotting ecosystem, showing the different plotting libraries available and their relation. Among those packages, we will focus on Plots and Makie—this chapter introduces the latter—and their backends. After reading this chapter, you should be able to choose the most appropriate plotting library for your data visualization task.

Chapter 3, Getting Interactive Plots with Julia, describes how to get interactive plots using Julia. First, this chapter will go through different kinds of interactivity, highlighting the packages that allow you to access them. Among those packages, it presents the Observables package, an essential package in the Julia ecosystem for developing interactive visualizations. Then, this chapter goes deeper into reaching interactivity with Plots and Makie packages and their backends. After that, it introduces the Interact and PlutoUI packages to create interactive visualizations using Jupyter and Pluto notebooks, respectively. Finally, it presents tools to publish and share interactive visualizations using dashboards, web pages, and simple graphical user interfaces.

Chapter 4, Creating Animations, teaches you how to create animations with Plots and Makie. Also, it will help you determine when it could be helpful to produce animations. Finally, this chapter introduces Javis, a drawing library to generate animations in Julia.

Chapter 5, Introducing the Grammar of Graphics, covers the Julia plotting packages based on or inspired by *The Grammar of Graphics*, namely, `Gadfly`, `VegaLite`, and

AlgebraOfGraphics. It will also introduce the DataFrames package, crucial for working with tidy data in Julia. Therefore, this chapter is essential for starting to analyze and visualize data.

Chapter 6, Creating Statistical Plots, shows you how to create the most common statistical plots using the different Julia libraries. In particular, this chapter will focus on Plots, Makie, and Gadfly. This chapter, together with the previous one, gives you an excellent primer to start performing data analysis and data visualization tasks with Julia.

Chapter 7, Visualizing Graphs, introduces the LightGraph and MetaGraphs packages to work with graphs and networks. Then, it shows us how to visualize them using GraphPlot, GraphRecipes, and GraphMakie.

Chapter 8, Visualizing Geographically Distributed Data, showcases how to visualize geographically distributed data with choropleth plots using Plots and GeoMakie. After that, the chapter will examine the creation of street maps using the OpenStreetMapX package.

Chapter 9, Plotting Biological Data, shows different packages that allow the visualization and analysis of biological data. In particular, the chapter focuses on visualizing phylogenetic trees, multiple sequence alignments, and protein structures. Finally, it shows how to create a dashboard using DashBio to develop interactive visualizations for biological data.

Chapter 10, The Anatomy of a Plot, explains a plot's different components. That knowledge is needed to understand the organization of the attributes for customizing those components. To that end, the chapter introduces the different terminology that Plots, Makie, and Gadfly use for their plot elements.

Chapter 11, Defining Plot Layouts to Create Figure Panels, describes how to compose multiple plots into one. To that end, it will first present the powerful Plots layout system. After that, it will introduce the layout capabilities of Makie and Gadfly.

Chapter 12, Customizing Plot Attributes – Axes, Legends, and Colors, covers some interesting attributes to customize your plots—focusing on the Plots package. The chapter first gives you the tools needed for exploring the Plots and Makie attributes without leaving your Julia session. Then, it shows you how to insert LaTeX equations into your Plots and Makie figures. After that, this chapter will teach us how to customize text elements, axes, legends, and colors.

Chapter 13, Designing Plot Themes, shows how to define plot themes—needed to reuse customizations—or use the predefined ones. In particular, this chapter will discuss theming in Plots, Makie, and Gadfly.

Chapter 14, Designing Your Own Plots – Plot Recipes, introduces the Plots and Makie recipe systems to create custom visualizations or whole new plot types. The chapter also shows how to draw shapes and create custom markers using Plots.

To get the most out of this book

Most of this book's chapters need a computer with Julia and a web browser installed and access to the internet. All the code examples have been tested on a Linux machine using Julia 1.6, as this was the LTS version when writing the book. Since the chapters were written at different times, package versions can change from chapter to chapter. Because of that, the following table offers version ranges for some packages:

Software/hardware covered in the book	Operating system requirements
Julia 1.6	Windows, macOS, or Linux
Pluto 0.15.1 - 0.19	Windows, macOS, or Linux
Plots 1.21.3 - 1.31	Windows, macOS, or Linux
CairoMakie 0.6 - 0.8	Windows, macOS, or Linux
GLMakie 0.4 - 0.6	Windows, macOS, or Linux
Gadfly 1.3.4	Windows, macOS, or Linux

This book touches on many packages, and this list only shows the ones used in most chapters. To ensure reproducibility, we provide the code used to create the book examples in the GitHub repository of this book. Each chapter folder contains the project environment used to write the examples; you can activate those environments as described in *Chapter 1, An Introduction to Julia for Data Visualization and Analysis,* if you find errors when using more recent versions of the packages. Also, each Pluto notebook saves its project environment inside; therefore, you can run the Pluto notebook files in the GitHub repo to ensure you are using identical package versions. Note that Julia packages follow the **Semantic Versioning (SemVer)** specification. Therefore, you can expect this code to work with any recent release of Plots version 1. However, note that for packages in version 0, any change in the middle number can be breaking.

If you are using the digital version of this book, we advise you to type the code yourself or access the code from the book's GitHub repository (a link is available in the next section). Doing so will help you avoid any potential errors related to the copying and pasting of code.

Note that this book requires you to have basic computer programming knowledge— experience with any high-level language will be enough. You do not need to know Julia, as the book will introduce the concepts and syntax required to follow the examples. However, note that this book will not be enough if you don't have experience with Julia and are looking to learn the language. In such a case, complementing this book with other sources could be beneficial. Also, note that this book focuses on visualization aspects and won't discuss the mathematical and statistical foundations behind plot types. Therefore, you will get the most out of some chapters in the book's second part if you have some basic knowledge of those subjects—but that is not a requirement to follow the text.

Download the example code files

You can download the example code files for this book from GitHub at `https://github.com/PacktPublishing/Interactive-Visualization-and-Plotting-with-Julia`. If there's an update to the code, it will be updated in the GitHub repository.

We also have other code bundles from our rich catalog of books and videos available at `https://github.com/PacktPublishing/`. Check them out!

Download the color images

We also provide a PDF file that has color images of the screenshots and diagrams used in this book. You can download it here: `https://packt.link/i4MaE`.

Conventions used

There are a number of text conventions used throughout this book.

`Code in text`: Indicates code words in text, database table names, folder names, filenames, file extensions, pathnames, dummy URLs, user input, and Twitter handles. Here is an example: "The `plotmap` function from `OpenStreetMapX` can take the `MapData` object and plot the stored map."

A block of code is set as follows:

```
plot([sin, cos], 0:0.1:2pi,
labels=["sin" "cos"],
linecolor=[:orange :green],
linewidth=[1, 5])
```

When we wish to draw your attention to a particular part of a code block, the relevant lines or items are set in bold:

```
plt = plot(data_block.geometry,
fill_z = permutedims(data_block.deaths),
```

```
colorbar_title = "cholera deaths",
seriescolor = :Greys_3,
linecolor = :darkgray,
framestyle = :none)
```

Any command-line input or output is written as follows:

```
julia script.jl
```

Bold: Indicates a new term, an important word, or words that you see onscreen. For instance, words in menus or dialog boxes appear in **bold**. Here is an example: "Click **New** in the **Edit environment variable** window."

> Tips or Important Notes
> Appear like this.

Get in touch

Feedback from our readers is always welcome.

General feedback: If you have questions about any aspect of this book, email us at customercare@packtpub.com and mention the book title in the subject of your message.

Errata: Although we have taken every care to ensure the accuracy of our content, mistakes do happen. If you have found a mistake in this book, we would be grateful if you would report this to us. Please visit www.packtpub.com/support/errata and fill in the form.

Piracy: If you come across any illegal copies of our works in any form on the internet, we would be grateful if you would provide us with the location address or website name. Please contact us at copyright@packt.com with a link to the material.

If you are interested in becoming an author: If there is a topic that you have expertise in and you are interested in either writing or contributing to a book, please visit authors.packtpub.com.

Share Your Thoughts

Once you've read *Interactive Visualization and Plotting with Julia*, we'd love to hear your thoughts! Scan the QR code below to go straight to the Amazon review page for this book and share your feedback.

https://packt.link/r/1-801-81051-6

Your review is important to us and the tech community and will help us make sure we're delivering excellent quality content.

Section 1 – Getting Started

In this section, you will learn the basics about Julia and its plotting ecosystem. We will start with a quick introduction to Julia for data analysis and visualization – no previous Julia experience is needed. Then, we will explore the different plotting packages in the Julia ecosystem. The initial chapters will introduce you to some basic plotting capabilities of `Plots` and *Makie*. Then, we will go deeper to see how to create interactive plots and animations with them. This section will also introduce us to Pluto notebooks, interactive and reproducible Julia notebooks that will be essential for following this book's code examples.

This section comprises the following chapters:

- *Chapter 1, An Introduction to Julia for Data Visualization and Analysis*
- *Chapter 2, The Julia Plotting Ecosystem*
- *Chapter 3, Getting Interactive Plots with Julia*
- *Chapter 4, Creating Animations*

1
An Introduction to Julia for Data Visualization and Analysis

Julia is a high-level, general-purpose language that offers a fresh approach to data analysis and visualization. Its clean syntax and high performance make Julia a language worth knowing for any data scientist.

This chapter will introduce the minimum set of concepts and techniques needed for data visualization in Julia. Therefore, we will explore Julia's essential tools for representing, analyzing, and plotting data in a reproducible way. If you are starting with Julia, this chapter is vital to you. Advanced Julia users wanting to learn about the `Plots` package can benefit from the last section of this chapter.

After reading this chapter, you will know how to set up a Julia reproducible environment for developing your data visualization tasks. You will know how to use Julia and the `Plots` library to create basic plots, particularly *heatmaps*, *scatter*, *bar*, and *line* plots.

In this chapter, we're going to cover the following main topics:

- Getting started with Julia
- Installing and managing packages
- Choosing a development environment
- Knowing the basic Julia types for data visualization
- Creating basic plots

Technical requirements

In this chapter, we will explore Julia in multiple ways. You will need a computer with an operating system and architecture supported by Julia; most 32-bit and 64-bit computers running recent versions of Linux, FreeBSD, Windows, or macOS are sufficient. You will also need a modern web browser, an internet connection, and **Visual Studio Code**. Once you have that, this chapter will guide you on the installation of Julia and the required packages.

Also, the code examples are available in the `Chapter01` folder of the following GitHub repository: `https://github.com/PacktPublishing/Interactive-Visualization-and-Plotting-with-Julia`. In particular, the `JuliaTypes.jl` Pluto notebook has the code examples for the *Knowing the basic Julia types for data visualization* section and `BasicPlots.jl` the examples for the *Creating a Basic Plot* section. You will find in the same folder the static HTML version of both notebooks with the embedded outputs.

Getting started with Julia

We are going to get started with Julia. This section will describe how to run Julia scripts and interact with the Julia **Read-Eval-Print Loop** (**REPL**), the command line or console used to execute Julia code interactively. But first, let's install Julia on your system.

Installing Julia

We are going to install the stable release of Julia on your system. There are many ways to install Julia, depending on your operating system. Here, we are going to install Julia using the official binaries. First, let's download the binary of Julia's **Long-Term Support** (**LTS**) release, *Julia 1.6*. You can install other releases, but I recommend using Julia 1.6 to ensure reproducibility of the book's examples:

1. Go to `https://julialang.org/downloads/`.

2. Download the binary file for the LTS release that matches your operating system (Windows, macOS, Linux, or FreeBSD) and the computer architecture bit widths (64-bit or 32-bit).

If you are a *Windows* user, we recommend downloading the installer. When choosing the installer, note that a 64-bit version will only work on 64-bit Windows.

If you are using *Linux*, you should also match the instruction set architecture of your computer processor (x86, ARM, or PowerPC). You can use the following command in your terminal to learn your architecture:

```
uname --all
```

Note that **i386** on the output means you have a 32-bit x86 processor, and **x86_64** indicates a 64-bit x86 processor.

The instructions that follow this step depend on your operating system, so, let's see them separately. Once you have finished, you can test your Julia installation by typing `julia -v` on your terminal; it should print the installed Julia version.

Linux

Installation on Linux is pretty simple; you should do the following:

1. Move the downloaded file to the place you want to install Julia.
2. Decompress the downloaded `.tar.gz` file. On a terminal, you can use `tar -xf` followed by the filename to do it.
3. Put the `julia` executable that is inside the `bin` folder of the decompressed folder on `PATH`. The `PATH` system variable contains a colon-separated list of paths where your operating system can find executables. Therefore, we need the folder containing the `julia` executable on that list to allow your operating system to find it. There are different ways to do that depending on the *Unix shell* that runs in your system terminal. We will give the instructions for *Bash*; if you are using another shell, please look for the corresponding instructions. You can know the shell that runs in your terminal by executing `echo $0`.

   Using Bash, you can add the full absolute path to the `bin` folder to `PATH` in `~/.bash_profile` (or `~/.bashrc`) by adding the following line to that file:

   ```
   export PATH="$PATH:/path/to/julia_directory/bin"
   ```

You should replace /path/to/julia_directory/bin with the full path to the bin folder containing the julia binary. The $ and : characters are required; it is easy to mess up your Bash terminal by forgetting them.

macOS

To install Julia on macOS, do the following:

1. Double-click the downloaded .dmg file to decompress it.

2. Drag the .app file to the **Applications** folder.

3. Put julia on PATH to access it from the command line. You can do that by creating a symbolic link to julia on /usr/local/bin – for example, if you are installing Julia 1.6, you would use this:

    ```
    ln -s /Applications/Julia-1.6.app/Contents/Resources/
    julia/bin/julia /usr/local/bin/julia
    ```

Windows

If you are on Windows, you should only run the installer and follow the instructions. Please note the address where Julia has been installed. You will need to put the julia executable on PATH to access Julia from the command line. The following instructions are for Windows 10:

1. Open **Control Panel**.

2. Enter **System and Security**.

3. Go to **System**.

4. Click on **Advanced system settings**.

5. Click on the **Environment variables...** button.

6. In the **User Variables** and **System Variables** sections, look for the Path variable and click that row.

7. Click the **Edit...** button of the section that has the Path variable selected.

8. Click **New** in the **Edit environment variable** window.

9. Paste the path to the julia executable.

10. Click the **OK** button.

Interacting with the Julia REPL

Now you have installed Julia, let's start exploring the Julia REPL or console. Let's do a simple arithmetic operation:

1. Type `julia` on your system terminal to open the Julia REPL. If you are using Windows, we recommend using a modern terminal as the Windows terminal.

2. Type `2 + 2` after the **julia>** prompt on the Julia REPL and press *Enter*.

    Voilà! You have done your first arithmetic operation in Julia.

3. Exit the Julia REPL using one of the following options:

    *   Type `exit()` and press *Enter*.
    *   Press the control key (*Ctrl*) together with the *D* key on a blank line.

Now that we know the basics of interacting with the Julia REPL, let's see some of its more exciting capabilities – *tab completion*, *support for Unicode characters*, and *REPL modes*.

Getting help from tab completion

The *Tab* key is handy when working on the Julia REPL. It can speed up your code input and show you the available options in a given context. When entering a name on a current workspace, pressing the *Tab* key will autocomplete the name until there are no ambiguities. If there are multiple options, Julia will list the possibilities.

You can also use tab completion to get information about the expected and keyword arguments that a function can take. To get that, you need to press the *Tab* key after the open bracket or after a comma in the argument list. For example, you can see the list of methods for the sum function, including possible arguments and their accepted types, by typing `sum(` and pressing the *Tab* key. Note that typing `sum([1, 2, 3],` and pressing the *Tab* key will give you a shorter list, as Julia now knows that you want to sum a vector of integers.

Last but not least, you can use tab completion to find out the field names of a given object. For example, let's see the field names of a range object:

1. Type `numbers = 1:10` on the Julia REPL and press *Enter*. That defines the variable numbers to contain a range of integers.

2. Type `numbers.` and press the *Tab* key. Julia will autocomplete `st`, as the two fields of a range object start with these two letters.

3. Press the *Tab* key. Julia will list the field names of the range object – in this case, `start` and `stop`.

4. Type a to disambiguate and press the *Tab* key to autocomplete.

5. Press the *Enter* key to see the value of `numbers.start`.

Julia can autocomplete more things, such as the fields of a function output or a dictionary's keys. As we will see next, you can also use tab autocompletion to enter Unicode characters.

Unicode input

Julia allows you to use any Unicode character, including an emoji, on your variables and function names. You can copy and paste those characters from any source. Also, you can insert them using tab completion of their LaTeX-like abbreviation, starting with the \ character. For example, you can enter π by typing \pi and pressing the *Tab* key.

REPL modes

Julia offers a variety of REPL modes to facilitate different tasks. You can enter the different modes by pressing a specific key just after the **julia>** prompt. The following figure shows you the built-in REPL modes that come with Julia:

```
help?> # Press the ? key to enter the Help mode

julia> # Julian mode

(@v1.6) pkg> # Press the ] key to enter the Pkg mode to manage packages and environments

shell> # Press the ; key to enter the Shell mode to execute Unix shell commands
```

Figure 1.1 – The REPL modes

Let's test that by using the most helpful mode, the help mode to access the documentation of the `split` function:

1. Press the ? key just after the **julia>** prompt; you will see that the prompt changes to **help?>**.

2. Type `split` and press *Enter*; you will see the function's documentation and automatically return to the **julia>** prompt.

Going back to the **julia>** prompt only requires pressing the *Backspace* key just after the prompt of the REPL mode. Another way to come back to the **julia>** prompt is by pressing the control (*Ctrl*) and *C* keys together. In this case, we didn't need the *Backspace* key to return to the **julia>** prompt as the help mode isn't sticky. However, the **shell** and **pkg** modes shown in *Figure 1.1* are sticky and require pressing the *Backspace* key to go out of them.

Running Julia scripts

Using the Julia REPL is helpful for interactive tasks and ephemeral code, but creating and running Julia scripts can be a better option in other situations. A Julia script is simply a text file with the jl extension containing Julia code. The easiest way to run a Julia script in your system terminal is by running the julia executable and giving the path to the script file as the first positional argument. For instance, if you want to run a Julia script named script.jl that is in the current folder, you can run the following line in the terminal:

```
julia script.jl
```

We have now installed Julia and learned how to run Julia code. In the next section, we will learn how to install Julia packages and manage project environments to ensure reproducibility.

Installing and managing packages

Julia has a built-in package manager that you can use by loading the Pkg module or through **pkg** mode of the Julia REPL. In this section, we will learn how to use it to install packages and manage project environments.

Installing Julia packages

Julia has an increasing number of registered packages that you can easily install using the built-in package manager. In this section, we will install the Plots library as an example. Let's install Plots using the add command from **Pkg** mode. This way of installing packages comes in handy when working on the Julia REPL:

1. Open Julia.
2. Enter Pkg mode by pressing the] key just after the **julia>** prompt.
3. Type add Plots after the **pkg>** prompt and press *Enter*.
4. Wait for the installation to finish; it can take some time.
5. Press the *Backspace* key to return to the **julia>** prompt.

Great! This has been easy, and you now have the `Plots` package installed. However, Pkg mode is only available to you when you are in the Julia REPL. But, if you want to install a Julia package from a non-interactive environment (for example, inside a Julia script), you will need to use the `add` function from the `Pkg` module. The `Pkg` module belongs to the Julia Standard Library. Thankfully, you don't need to install the packages of the Standard Library before using them. Let's try adding `Plots` again but using the `Pkg` module this time. As we have already installed the latest version of `Plots`, this will be fast:

1. Open Julia.
2. Import the `Pkg` module by typing `import Pkg` after the **julia>** prompt and pressing *Enter*.
3. Type `Pkg.add("Plots")` and press *Enter*.

In the last example, we used `import` to load the `Pkg` module. In the next section, we will learn some different ways in which you can load packages.

Loading packages

We need to load a package to use it within a Julia session. There are two main ways to load packages in Julia. The first one is using the `import` keyword followed by the module name – for example, `import Pkg`. As Julia packages export modules of the same name, you can also use the name of a package. When we use `import` in this way, Julia only brings the package's module into scope. Then, we need to use qualified names to access any function or variable from that module. A **qualified name** is simply the module name followed by a dot and the object name – for example, `Pkg.add`.

The second way to load a package is to use the `using` keyword followed by the module name. Julia will bring the module into scope, as well as all the names that the module exports. Therefore, we do not need to use qualified names to access their functions and variables. For instance, executing `using Plots` will bring the exported `plot` function into scope. You can still access *unexported* functions and variables using their qualified names.

Managing environments

A **project environment** defines a set of package dependencies, optionally with their versions. The Julia package manager has built-in support for them, and we can use them to create reproducible data analysis pipelines and visualizations. For example, this book's code examples use environments to allow the reproducibility of code through time and across different systems.

Julia defines project environments using two files. The first is the `Project.toml` file that stores the set of dependencies. The second is the `Manifest.toml` file that stores the exact version of all the packages and their dependencies. While the former is mandatory for any environment, the latter is optional. Luckily, we do not need to create those files manually, as we can manage the environment's packages through the package manager.

There are multiple options for dealing with project environments. One is to start `julia` in a given environment by using the `--project` argument. Usually, we want to create an environment in the folder where we are starting Julia. In those cases, we can use a *dot* to indicate the current working directory. Let's create an environment in the current folder containing a specific version of the `Plots` package:

1. Run `julia --project=.` in the terminal to open the Julia REPL using the project environment defined in the current working directory.
2. Press the] key to enter Pkg mode. You will see the name of the current folder on the prompt. That's Pkg mode telling you that you are in that environment.
3. Type `status` and press *Enter* to see the content of your current environment.
4. Type `add Plots@1.0.0` to install Plots version 1.0.0 in that environment.
5. Run the `status` command in Pkg mode again to check what is in the environment after the previous operation.
6. Press the *Backspace* key to return to the Julia prompt.

In the previous example, we have used the `--project` argument to start `julia` in a particular environment. If you run `julia` without indicating a project folder, you will use the default environment corresponding to your Julia version.

You can change between environments using the `activate` command. It takes the path to the project folder that contains the environment, and if you do not give any path, Julia will start the default environment of your Julia version. For example, executing `activate .` in Pkg mode will start the environment in the current working directory, and running `activate` will return the default Julia environment.

When you first activate a non-empty environment on your system, you must install all the required packages. To get all the needed packages, you should run the `instantiate` command of Pkg mode. For example, instantiation will be necessary if you want to use an environment created on another computer.

While Pkg mode of the Julia REPL is handy when you are working interactively, sometimes you need to manage environments inside a Julia script. In those cases, the `Pkg` module will be your best friend. So, let's create a Julia script that uses a particular version of Plots. First, create a file named `installing_plots.jl` with the following content, using any text editor:

```
import Pkg
Pkg.activate(temp=true)
Pkg.add(Pkg.PackageSpec(name="Plots", version="1.0.0"))
Pkg.status()
```

In that code, we are using the `activate` function of the `Pkg` module with `temp=true` to create and activate the script environment in a temporary folder. We need to use the `PackageSpec` type defined on the `Pkg` module to add a specific package version.

Now, you can run the script executing `julia installing_plots.jl` on your terminal. You will see that the script creates and activates a new environment in a temporal folder. Then, it installs and precompiles `Plots` and its dependencies. Finally, it shows that `Plots` version 1.0.0 was installed in the environment. The script will run a lot faster the second time because the packages are installed and precompiled on your system.

There are other `Pkg` commands and functions that you will find helpful when managing environments – `status`, to list the packages on the current project environment, `update`, and `remove`. You can see the complete list of `Pkg` commands in Pkg mode by typing ? and pressing the *Enter* key just after the **pkg>** prompt. Optionally, you can see extended help for each command by entering ? and the command name in Pkg mode. If you are using the `Pkg` module, you can access the documentation of the functions by typing `Pkg.` and the function name in **help** mode of the Julia REPL.

Now that we know how to install and manage Julia packages, let's start installing some packages to set up the different development environments we will use for Julia.

Choosing a development environment

There are multiple options for developing using the Julia language. The choice of development environment depends on the task at hand. We will use three of them in this book – one **Integrated Development Environment (IDE)** and two notebooks. We generally use the IDE to write scripts and develop packages and applications. The **notebooks** allow us to perform exploratory and interactive data analysis and visualization. In this section of the book, we are going to introduce those development environments. Let's start with the IDE.

The Julia extension for VS Code

The official IDE for Julia is the Julia extension for **VS Code**. It provides a way to execute Julia code, search documentation, and visualize plots among many utilities. Describing all its features goes beyond the scope of this book. In this section, you will learn how to run Julia code on VS Code, but first, let's install the IDE.

Installing Julia for VS Code

To install and use the Julia extension, you will need Julia installed on your system and the `julia` executable on PATH. Then, you should install VS Code from `https://code.visualstudio.com/`. Once you have VS Code installed, you can install the Julia extension from it:

1. Click on the **View** menu and then on **Extensions** to open **Extensions View**.
2. Type `julia` in the search box at the top of **Extensions View**.
3. Click on the **Install** button of the **Julia** extension provided by **julialang**.
4. Restart VS Code once the installation has finished.

Let's now use our Julia IDE to run some code.

Running Julia on VS Code

There are multiple ways to run code using the Julia extension on VS Code. Here, we will run code blocks by pressing the *Alt* and *Enter* keys together (*Alt + Enter*) or entire files by pressing *Shift + Enter*. Let's test that with a simple Julia script:

1. Click on the **File** menu and then on the **New File** option; VS Code creates a new empty file.
2. Click on **File** and select the **Save** option.
3. Choose a location and name for your file using the `jl` extension (for example, `first_script.jl`) and click the **Save** button. The `jl` extension is crucial, as it will indicate to VS Code that you are coding in Julia.
4. Click on the **JULIA** button on the sidebar; it is the one with three dots, which you can see selected in *Figure 1.2*. It opens the workspace, documentation, and plot navigator panes of the Julia IDE.
5. Click on the first line of the empty file and type `a = 2 + 2`.

6. Press together the *Alt* and *Enter* keys; this will execute the code block. You will see the operation's output *inlined* on the file, just after the executed code block. Note that this can take some time the first time, as the Julia extension should precompile the `VSCodeServer` package. The Julia extension will open a Julia REPL, showing the status of the package precompilation and the result of the executed code block. After this first execution, you can also see the recently assigned a variable in the workspace pane.

7. Press *Enter* to create a new line and type `println("Hello World")`.

8. Press *Alt + Enter* on that line to execute that code block. Now, the Julia extension only *inlines* a checkmark to indicate that Julia successfully ran the `println` function. The Julia extension uses a checkmark when the function returns `nothing`. The `println` function, by default, prints to the standard output and returns nothing. In this case, you can see **Hello World** printed in the Julia REPL. There is a Julia variable named `ans` when you run Julia interactively that holds the last object returned. You can see on the **WORKSPACE** pane that the `ans` variable contains `nothing` (see *Figure 1.2*).

9. Press the *Shift* and *Enter* keys together to execute the whole file. The Julia extension only *inlines* the output of the last expression of the executed file – in this case, the checkmark. Also, you will see that Julia did not print the result of the first expression on the Julia REPL. If you want to show a value when running a file, you need to print it explicitly, as we did with the `"Hello World"` string.

10. Save the changes by pressing the *Ctrl* and *S* keys together (*Ctrl + S*).

After finishing that process, your VS Code session will look similar to the one in the following figure:

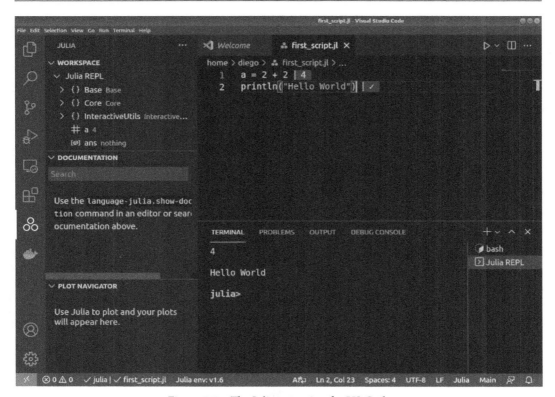

Figure 1.2 – The Julia extension for VS Code

At this point, you have set up VS Code to work with Julia on your computer, and you have learned how to run Julia code on it. This will come in handy when developing Julia scripts, packages, and applications. Also, you can now use VS Code anytime a text editor is required throughout this book. In the following section, we will learn how to execute code using Julia notebooks.

Using Julia notebooks

Notebooks are an excellent way to do literate programming, embedding code, Markdown annotations, and results into a single file. Therefore, they help to share results and ensure reproducibility. You can also code interactively using Julia notebooks instead of the Julia REPL. There are two main notebooks that you can use for Julia – **Jupyter** and **Pluto**. We are going to describe them in the following sections.

Taking advantage of Jupyter notebooks through IJulia

Jupyter notebooks are available for multiple programming languages. In Julia, they are available thanks to the IJulia package. Let's create our first Jupyter notebook:

1. Copy and paste the following code block into the Julia REPL:

    ```
    import Pkg
    Pkg.add("IJulia")
    ```

    This will install the IJulia package needed for using *Jupyter* with Julia.

2. Type using IJulia in the Julia REPL and press *Enter* to load the package.

3. Now, type notebook(dir=".") and press *Enter* to open Jupyter. Because we set the dir keyword argument to ".", Jupyter will use the same *current working directory* as the Julia REPL. If this is the first time that you have run the notebook function, Julia will ask you to install *Jupyter* using the Conda package. You can press *Enter* to allow Julia to install Jupyter using a *Miniconda* installation private to Julia. Note that the process is automatic, but it will take some time. If you prefer using an already installed Jupyter instance instead, please read the IJulia documentation for instructions. Once Jupyter is installed, the notebook function of the IJulia package will open a tab in your web browser with Jupyter on it.

4. Go to the browser tab that is running the Jupyter frontend.

5. Click the **New** button on the right and select the Julia version you are using; this will open a new tab with a Julia notebook named **Untitled** stored in the current directory with the name Untitled.ipynb.

6. Click on **Untitled** and change the name of the notebook – for example, if you rename the notebook to FirstNotebook, Jupyter renames the file to FirstNotebook.ipynb.

7. Click on the empty cell and type a = 2 + 2, and then press *Shift + Enter* to run it. You will see that Jupyter shows the output of the expression just after the cell. The cursor moves to a new cell; in this case, as there was no cell, Jupyter creates a new one below. Note that Jupyter keeps track of the execution order by enumerating the cells' inputs (**In**) and outputs (**Out**).

8. Type println("Hello World") in the new cell and press *Shift + Enter* to run that code. You will see the output of the println function below the cell code. As println returns nothing, there is no numbered output for this cell (see *Figure 1.3*).

Jupyter supports Markdown cells to introduce formatted text, images, tables, and even LaTeX equations in your notebooks. To create them, you should click on an empty cell and then on the drop-down menu that says **Code** and select the **Markdown** option. You can write Markdown text on that cell, and Jupyter will render it when you run it (*Shift + Enter*). You can see an example of running **Markdown** *cell* in the following figure:

Figure 1.3 – A Jupyter notebook using Julia

In these examples, we have used one line of code for each cell, but you can write as many lines of code as you want. However, if there are multiple code blocks inside a cell, Jupyter only shows the output of the last expression. You can *suppress the output* of a cell by ending it with a *semicolon*. That trick also works on the Julia REPL and in Pluto notebooks.

Finally, you can close Jupyter by going back to the Julia terminal that runs it and pressing *Ctrl + C*. Jupyter autosaves the changes every 120 seconds, but if you want to save changes manually before exiting, you need to click on the save icon, the first on the toolbar, or press *Ctrl + S*. Now that we've had our first experience with Jupyter, let's move on to Pluto.

Using Pluto notebooks

Pluto notebooks are only available for the Julia language, and they differ from Jupyter notebooks in many aspects. One of the most important is that Pluto notebooks are **reactive**; that means that changing the code of one cell can trigger the execution of the dependent cells. For example, if one cell defines a variable and the second cell does something with its value, changing the variable's value in the first one will trigger the re-execution of the second. To install Pluto, you only need to install the Pluto package. Once you have installed it, let's create a new Pluto notebook:

1. Type import Pluto on the Julia REPL and press *Enter*.

2. Type `Pluto.run()` and press *Enter*. The `run` function will open Pluto in a tab in your web browser.

3. Click on the **New notebook** link. It will redirect you to an empty Pluto notebook.

4. Click on **Save notebook...** at the top middle and type a name for the notebook; it should have the `jl` extension – for example, `FirstNotebook.jl`.

5. Press *Enter* or click the **Choose** button. Pluto will create a new notebook file on the indicated path.

6. Select the empty cell, type `name = "World"`, and press *Ctrl + Enter*. This will run the cell and add a new cell below. You will see that Pluto shows the cell output over the cell code.

7. Type `"Hello $name"` in the new cell and press *Shift + Enter* to execute the cell. You will see the **"Hello World"** string appear over the cell. The executed code creates a string by interpolating the value of the `name` variable.

8. Double-click on the word **World** in the first cell to select it and type the word `Pluto`.

9. Press *Shift + Enter* to run the cell, changing the value of the name from the `"World"` string to `"Pluto"`. You will see that Pluto automatically executes the last cell, changing its output to **"Hello Pluto"**.

Excellent! You have now had a first taste of what a reactive notebook is. Let's see how we can create a Markdown cell in Pluto:

1. Hover the mouse over the last cell; you will see that two plus symbols, +, appear over and below the cell (see *Figure 1.4*).

2. Click on the + symbol below the last cell to create a new cell below.

3. Type `md"**Markdown** *cell*"` into the new cell and press *Shift + Enter* to run the cell and render the Markdown string. `md""` created a single-line Markdown string.

4. Hover the mouse over the cell that contains the Markdown string; you will see an eye symbol that appears on the left of the cell.

5. Click on the eye button to hide the cell code; note that Pluto has crossed out the eye icon.

There you have it! As you can see, a Markdown cell in Pluto is simply a cell that contains a Markdown string and for which we have decided not to show its source code. You can see the result of this process in the next figure:

Figure 1.4 – A Pluto notebook

Pluto also differs from Jupyter by the fact that each cell should preferably do one thing. Therefore, if you plan to write multiple lines inside a single cell, they must belong to a single code block – for example, you can use multiple lines inside a function body, or a block defined between the `begin` and `end` keywords. Another difference is that Pluto doesn't allow the definition of the same variable name, nor the load of the same module in multiple cells.

Pluto notebooks are Julia files using the `jl` extension. You can read a Pluto notebook like any other Julia file and run it as a script. Most of the things encoding for notebook-specific aspects are just comments on the code. The outputs and figures are not stored in the file but recreated each time we open a notebook.

The best way to conserve and share code and results is to export the static HTML page. You can achieve that by clicking on the export button, a triangle over a circle located in the top-right corner of the notebook, and selecting the **Static HTML** option. The exported HTML also has the Julia notebook file encoded inside. This allows you to download the notebook using the **Edit or run this notebook** button that appears on the downloaded HTML document. There, you need to click on the **notebook.jl** link in the **Download the notebook** item in the **On your computer** section to download the notebook file to your machine. Then, you follow the instructions in the **On your computer** section to open the notebook using Pluto.

One unique aspect of Pluto notebooks that ensures reproducibility is that the notebook file also stores the project environment of the notebook. Pluto manages the notebook environment depending on the `import` and `using` statements. For example, loading a package will automatically install it in the notebook environment. If you need it, you can use the `activate` function of the `Pkg` module to disable that feature and manage the project environment yourself.

We have not used the `println` function in the Pluto examples because Pluto doesn't show things printed to `stdout` on the notebook. Instead, you can find the printed elements on the Julia REPL that is running Pluto. If you need to print something in the notebook, you will need the `Print` function from the `PlutoUI` package.

Pluto has a documentation panel that can show you the docstrings of the function and objects you are using. To open it, you need to click on the **Live Docs** button at the bottom right. Then, if you click on an object, for example, the `Print` function, you will see its documentation on the panel.

Pluto saves changes on a cell every time it runs. A *Ctrl + S* button at the top right will remind you to keep any unsaved changes. You can click on it or press *Ctrl + S* to save the changes. To close Pluto, do the following:

1. Click on the **Pluto.jl** icon to return to the Pluto main page.
2. Click on the dark **x** button at the side of the open notebooks in the **Recent sessions** section to close the notebooks.
3. Close the open Pluto tabs of your web browser.
4. Go to the Julia terminal that is running Pluto and press *Ctrl + C* to close it.

Now you know the basics of working with the three primary development environments for Julia.

Running examples using prompt pasting

Julian mode of the Julia REPL and Pluto notebooks has a nice feature called **prompt pasting**. It means that you can copy and paste Julia code examples, including the **julia>** prompt and the outputs. Julia will strip out those, leaving only the code for its execution.

In the Julia REPL, you need to paste everything after the **julia>** prompt; you will see that Julia automatically extracts and executes the code. In Pluto, click on a cell and paste it. You will see that Pluto pastes the code on new cells under the selected cell without running it. You should note that prompt pasting doesn't work on the standard Windows Command Prompt, nor in Jupyter.

Now you know that if you see *code examples* starting with the **julia>** prompt in this book or elsewhere, you can copy and paste them in the Julia REPL or Pluto to execute them.

Knowing the basic Julia types for data visualization

This section will explore the Julia syntax, objects, and features that will help us perform data analysis and visualization tasks. Julia and its ecosystem define plenty of object types useful for us when creating visualizations. Because user-defined types are as fast as built-in types in Julia, you are not constrained to using the few structs defined in the language. This section will explore both the built-in and package-defined types that will help us the most throughout this book. But first, let's see how to create and use Julia functions.

Defining and calling functions

A **function** is an object able to take an input, execute some code on it, and return a value. We have already called functions using the **parenthesis syntax** – for example, `println("Hello World")`. We pass the function input values or arguments inside the parentheses that follow the function name.

We can better understand Julia's functions by creating them on our own. Let's make a straightforward function that takes two arguments and returns the sum of them. You can execute the following code in the Julia REPL, inside a new Julia script in VS Code, or into a Jupyter or Pluto cell:

```
function addition(x, y)
    x + y
end
```

We can use the `function` keyword to create a function. We then define its name and declare inside the parentheses the positional and keyword arguments. In this case, we have defined two positional arguments, x, and y. After a new line, we start the function body, which determines the code to be executed by the function. The function body of the previous code contains only one expression – x + y. A Julia function always returns an object, and that object is the value of the last executed expression in the function body. The `end` keyword indicates the end of the function block.

We can also define the function using the **assignation syntax** for simple functions such as the previous one, containing only one expression – for example, the following code creates a function that returns the value of subtracting the value of y from x:

```
subtraction(x, y) = x - y
```

A Julia function can also have **keyword arguments** defined by name instead of by position – for example, we used the `dir` keyword argument when we executed `notebook(dir=".")`. When declaring a function, we should introduce the keyword arguments after a semicolon.

Higher-order functions

Functions are first-class citizens in the Julia language. Therefore, you can write Julia functions that can have other functions as inputs or outputs. Those kinds of functions are called **higher-order functions**. One example is the `sum` function, which can take a function as the first argument and apply it to each value before adding them. You can execute the following code to see how Julia takes the absolute value of each number on the vector before adding them:

```
sum(abs, [-1, 1])
```

In this example, we used a named function, `abs`, but usually, the input and output functions are anonymous functions.

Anonymous functions

Anonymous functions are simply functions without a user-defined name. Julia offers two foremost syntaxes to define them, one for single-line functions and the others for multiline functions. The single-line syntax uses the `->` operator between the function arguments and its body. The following code will perform `sum(abs, [-1, 1])`, using this syntax to create an anonymous function as input to `sum`:

```
sum(x -> abs(x), [-1, 1])
```

The other syntax uses the `do` block to create an anonymous function for a higher-order function, taking a function as the first argument. We can write the previous code in the following way using the `do` syntax:

```
sum([-1, 1]) do x
    abs(x)
end
```

We indicated the anonymous function's arguments after the do keyword and defined the body after a new line. Julia will pass the created anonymous function as the first argument of the higher-order function, which is sum in this case.

Now that we know how to create and call functions, let's explore some types in Julia.

Working with Julia types

You can write Julia types using their **literal representations**. We have already used some of them throughout the chapter – for example, 2 was an integer literal, and "Hello World" was a string literal. You can see the type of an object using the typeof function – for example, executing typeof("Hello World") will return String, and typeof(2) will return Int64 in a 64-bit operating system or Int32 in a 32-bit one.

In some cases, you will find the dump function helpful, as it shows the type and the structure of an object. We recommend using the Dump function from the PlutoUI package instead, as it works in both Pluto and the Julia REPL – for example, if we execute numbers = 1:5 and then the Dump(numbers) integer literal, we will get the following output in a 64-bit machine:

```
UnitRange{Int64}
   start: Int64 1
   stop: Int64 5
```

So, Dump shows that 1:5 creates UnitRange with the start and stop fields, each containing an integer value. You can access those fields using the **dot notation** – for example, executing numbers.start will return the 1 integer.

Also, note that the type of 1:5 was UnitRange{Int64} in this example. UnitRange is a parametric type, for which Int64 is the value of its type parameter. Julia writes the type parameters between brackets following the type name.

Julia has an advanced type system, and we have learned the basics to explore it. Before learning about some useful Julia types, let's explore one of the reasons for Julia's power – its use of multiple dispatch.

Taking advantage of Julia's multiple dispatch

We have learned how to write functions and to explore the type of objects in Julia. Now, it's time to learn about methods. The functions we have created previously are known as **generic functions**. As we have not annotated the functions using types, we have also created **methods** for those generic functions that, in principle, can take objects of any type. You can optionally add type constraints to function arguments. Julia will consider this type annotation when choosing the most specific function method for a given set of parameters. Julia has **multiple dispatch**, as it uses the type information of all positional arguments to select the method to execute. The power of Julia lies in its multiple dispatch, and plotting packages take advantage of this feature. Let's see what multiple dispatch means by creating a function with two methods, one for strings and the other for integers:

1. Open a Julia REPL and execute the following:

    ```
    concatenate(a::String, b::String) = a * b
    ```

    This code creates a function that concatenates two string objects, a and b. Note that Julia uses the * operator to concatenate strings. We need to use the :: operator to annotate types in Julia. In this case, we are constraining our function to take only objects of the String type.

2. Run concatenate("Hello", "World") to test that our function works as expected; it should return "HelloWorld".

3. Run methods(concatenate) to list the function's methods. You will see that the concatenate function has only one method that takes two objects of the String type – concatenate(a::String, b::String).

4. Execute concatenate(1, 2). You will see that this operation throws an error of the MethodError type. The error tells us that there is **no method matching concatenate(::Int64, ::Int64)** if we use a 64-bit machine; otherwise, you will see **Int32** instead of **Int64**. The error is thrown because we have defined our concatenate to take only objects of the String type.

5. Execute concatenate(a::Int, b::Int) = parse(Int, string(a) * string(b)) to define a new method for the concatenate function taking two objects of the Int type. The function converts the input integer to strings before concatenation using the string function. Then, it uses parse to get the integer value of the Int type from the concatenated strings.

6. Run methods(concatenate); you will see this time that concatenate has two methods, one for String objects and the other for integers.

7. Run concatenate(1, 2). This time, Julia will find and select the concatenate method taking two integers, returning the integer 12.

Usually, converting types will help you to fix MethodError. When we found the error in *step 4* we could have solved it by converting the integers to strings on the call site by running concatenate(string(1), string(2)) to get the "12" string. There are two main ways to *convert objects* in Julia explicitly – the first is by using the convert function, and the other is by using the type as a function (in other words, calling the type constructor). For example, we can convert 1, an integer, to a floating-point number of 64 bits of the Float64 type using convert(Float64, 1) or Float64(1) – which option is better will depend on the types at hand. For some types, there are special conversion functions; strings are an example of it. We need the string function to convert 1 to "1", as in string(1). Also, converting a string containing a number to that number requires the parse function – for example, to convert "1" to 1, we need to call parse(Int, "1").

At this point, we know the basics for dealing with Julia types. Let's now explore Julia types that will help us create nice visualizations throughout this book.

Representing numerical values

The most classic numbers that you can use are integers and floating-point values. As Julia was designed for scientific computing, it defines number types for different word sizes. The most used ones are Float64, which stores 64-bit floating-point numbers, and Int. This is an alias of Int64 in 64-bit operating systems or Int32 in 32-bit architectures. Both are easy to write in Julia – for example, we have already used Int literals such as -1 and 5. Then, each time you enter a number with a *dot*, *e*, or *E*, it will define Float64 – for example, 1.0, -2e3, and 13.5E10 are numbers of the Float64 type. The *dot* determines the location of the decimal point and *e* or *E* the exponent. Note that .1 is equivalent to 0.1 and 1. is 1.0, as the zero is implicit on those expressions. Float64 has a value to indicate something that is not a number – NaN. When entering numbers in Julia, you can use _ as a digit separator to make the number more legible – for example, 10_000. Sometimes, you need to add units to a number to make it meaningful. In particular, we will use mm, cm, pt, and inch from the Measures package for plotting purposes. After loading that package, write the number followed by the desired unit, for example, 10.5cm. That expression takes advantage of Julia's **numeric literal coefficients**. Each time you write a numeric literal, such as 10.5, just before a Julia parenthesized expression or variable, such as the cm object, you imply a multiplication. Therefore, writing 10.5cm is equivalent to writing 10.5 * cm; both return the same object.

Representing text

Julia has support for single and multiline string literals. You can write the former using double quotes (") and the latter using triple double quotes ("""). Note that Julia uses single quotes (') to define the literal for single characters of the Char type.

Julia offers other kinds of strings that will be useful for us when creating plots and interactive visualizations – *Markdown*, *HTML*, and *LaTeX strings*. The three of them use Julia's string macros, which you can write by adding a short word before the first quotes – md for Markdown, html for HTML, and L for LaTeX. You will need to load the Markdown standard library to use the `md` string macro and the `LaTeXStrings` external package for the `L` string macro. Note that Pluto automatically loads the `Markdown` module, so you can use `md"..."` without loading it. Also, Pluto renders the three of them nicely:

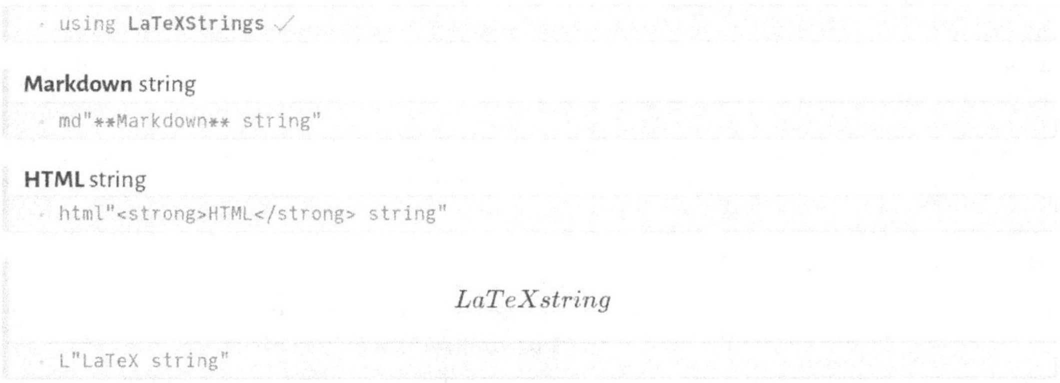

```
· using LaTeXStrings ✓
```

Markdown string
```
· md"**Markdown** string"
```

HTML string
```
· html"<strong>HTML</strong> string"
```

$$LaTeX string$$

```
· L"LaTeX string"
```

Figure 1.5 – Pluto rendering Markdown, HTML, and LaTeX strings

There is another type associated with text that you will also find in Julia when plotting and analyzing data – **symbols**. Symbols are interned strings, meaning that Julia stores only one copy of them. You can construct them using a colon followed by a word that should be a valid Julia variable name – for example, `:var1`. Otherwise, if it is not a valid identifier, you should use `String` and call the `Symbol` constructor – for example, `Symbol("var1")`.

Working with Julia collections

We will use two main collection types for data analysis and visualization – tuples and arrays. Julia collections are a broad topic, but we will explore the minimum necessary here. Let's begin with tuples. **Tuples** are immutable lists of objects of any type that we write between parentheses – for example, `("x", 0)` and `(1,)` are two- and one-element tuples respectively. Note that tuples with one element need the trailing comma.

Arrays can have multiple dimensions; the most common are vectors (one-dimensional arrays) and matrices (two-dimensional arrays). An array is a parametric type that stores the type of elements it contains and the number of dimensions as type parameters. We can construct an array using square brackets – for example, `[1]` and `[1, 2]` are vectors with one and two elements respectively. You can also write matrices using square brackets by separating columns with spaces, rather than commas and rows with semicolons or newlines – for example, `[1 2; 3 4]` is a *2 x 2* matrix.

For arrays, there are also other helpful constructors – `zeros`, `ones`, and `rand`. The three of them take the number of elements to create in each direction – for example, `zeros(4, 2)` will create a matrix full of zeros with four rows and two columns, while `zeros(10)` will create a vector of 10 zeros.

Julia offers the colon operator for creating a range of numbers; you can think of them as lazy vectors – for example, we have already seen `1:5` in a previous example. You can collect the elements of a range into a vector using the `collect` function.

You can *index* ranges, tuples, and arrays using the squared brackets syntax. Note that Julia has one-based **indexing**, so the first element of `collection` will be `collection[1]`.

You can also *iterate* over ranges, tuples, and arrays. There are two compact ways to iterate through those collections, apply a function, and get a new array. One is using **array comprehension** – for example, `[sqrt(x) for x in 1:5]`. Note that comprehension can also have filter expression – for example, if we want the square root only of odd numbers between 1 and 10, we can write `[sqrt(x) for x in 1:10 if x % 2 != 0]`.

The other compact way to apply a function over each collection element is to use **broadcasting**. Julia's broadcasting allows applying a function element-wise on arrays of different sizes by expanding singleton dimensions to match their sizes. It also enables operations between scalars and collections. Furthermore, you can use broadcasting to apply a function to a single collection. Note that for each function acting on scalars, Julia doesn't define methods taking collections of them. Therefore, Julia's preferred way to apply such a function to each collection element is to use the **dot syntax** for broadcasting. You only need to add a dot between the function name and the parentheses on a function call or a dot before an operator – for example, `sqrt.(collection) .+ 1`. Julia fuses the operations when using this syntax, so the square root and the addition happen in a single iteration.

We have now learned how to work with Julia types, particularly to represent text and numbers and their collections. In the next section, we will use them to create some plots.

Creating a basic plot

In the previous sections, we have learned some essentials about Julia. In this last section, we will learn how to use Julia for the creation of basic plots. For now, we will use the `Plots` package and its default backend, GR, but we are going to explore more deeply the Julia plotting ecosystem in the next chapter.

Let's start exploring the `Plots` syntax by creating a line plot, the default plot type. Line plots represent a series of related points by drawing a straight line between them.

The `plot` function of the `Plots` package can take different inputs. Plots usually take data from the positional arguments and **attributes** that modify the plot in the keyword arguments. The most common way to pass a series of data points is by giving their coordinates using two different vectors or ranges, one for *x* and the other for *y*. Let's do our first plot; you can choose whatever development environment you want to follow these steps:

1. Let's create some data by running the following code in the Julia REPL:

    ```
    x = 0:10
    y = sqrt.(x)
    ```

2. Run using Plots to load the Plots package.

3. Execute plot(x, y) to create your first line plot. Depending on the development environment, the plot will appear in different ways – in a new window for the Julia REPL, in the plot pane for VS Code, or inline inside the notebook for Jupyter and Pluto. You will see a plot like the one in the following figure:

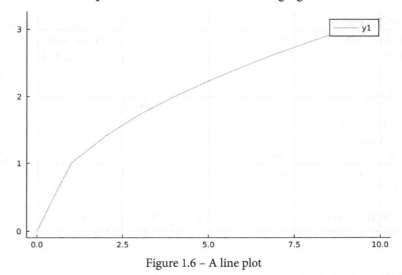

Figure 1.6 – A line plot

Great, you now have your first Julia plot! It is nice, but as we only took a few points from the sqrt function, the line has some sharp edges, most noticeably around *x* equal to one. Thankfully, Plots offers a better way to plot functions that adapts the number of points based on the function's second derivative. To plot a function in this way, you only need to give the function as the first argument and use the second and third positional arguments to indicate the initial and last values of *x* respectively – for example, to create a smooth line, the previous example becomes the following:

```
plot(sqrt, 0, 10)
```

Note that you can use your *x* coordinates by providing them as the second positional argument. That avoids calculating the optimal grid, so `plot(sqrt, x)` creates a plot identical to the first one shown in *Figure 1.6*.

If you give two functions as the first arguments and a domain or vector, `Plots` will use the latter as input for each function, and the first function will calculate the coordinates of *x* and the second function the coordinates of *y* – for example, you can define a *unit circle* using an angle in radians, from zero to two times pi, by defining *x* as the cosine of the angle and *y* as its sine:

```
plot(cos, sin, 0, 2pi, ratio=:equal)
```

Note that this code uses the `ratio` keyword argument, to ensure that we see a circle. Also, we have used Julia's numeric literal coefficient syntax to multiply 2 by the `pi` constant. The resulting plot is as follows:

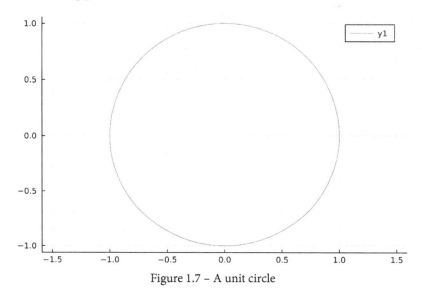

Figure 1.7 – A unit circle

In the last example, we indicated the limits of the domain, but as we said, we can also use a vector or range. For instance, try running the following:

```
angles = range(0, 2pi, length=100)
plot(cos, sin, angles, ratio=:equal)
```

In this case, we created a range using the `range` function to indicate the number of points we want in the plot with the `length` keyword argument.

We have just seen multiple ways to plot a single line, from specifying its points to using a function to let it determine them. Let's now see how to create a single plot with various lines.

Plotting multiple series

In the previous examples, we have plotted only one **data series** per plot. However, `Plots` allows you to superpose multiple series with different attributes into each plot. The main idea is that each column, vector, range, or function defines its series. For example, let's create a plot having two series, one for the `sin` function and the other for the `cos` function, using these multiple ways:

1. Define the values for the *x* axis, running `X = range(0, 2pi, length=100)`.

2. Execute `plot([sin, cos], X)`. Here, we have used a vector containing the two functions as the first argument. Each function on the vector defines a series with different labels and colors. Note that both series use the same values for the *x* axis.

3. Run `plot(X, [sin.(X), cos.(X)])`. You will get the same plot; however, we have used different inputs. The first positional argument is the range that indicates the coordinates for *x*. The second argument is a vector of vectors, as `sin.(X)`, for example, uses the dot broadcasting syntax to return a vector, with the result of applying the `sin` function to each element of `X`.

4. Execute the following commands:

    ```
    Y = hcat(sin.(X), cos.(X))
    plot(X, Y)
    ```

Note that `Y` is now a matrix with 100 rows and 2 columns. We are using the `hcat` function to concatenate the two vectors resulting from the broadcasting operations. As we said, each column defines a series. The resulting plot appears in the following figure and should be identical to the previous ones:

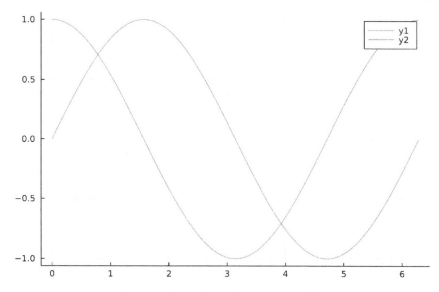

Figure 1.8 – A plot of the two data series

In Plots, each column defines a series, as in the last example. When one dimension represents multiple series, Plots repeats the dimension, having only one vector or range to match the series. That's the reason why we didn't need a matrix for *x* also in those examples.

Let's see how to apply different attributes to each series. In Plots, attributes indicated as vectors apply to a single series, while those defined through matrices apply to multiple ones – for example, the following code creates the plot in *Figure 1.9*:

```
plot([sin, cos], 0:0.1:2pi,
labels=["sin" "cos"],
linecolor=[:orange :green],
linewidth=[1, 5])
```

Here, we are using the *x*-axis domain values from 0 to 2pi, with a step distance of 0.1 units. ["sin" "cos"] defines a matrix with one row and two columns, as spaces rather than commas separate the elements. We can see in *Figure 1.9* that the labels attribute has assigned, for example, the string on the first column as the label of the first series. The same happens with linecolor, as we have also used a two-column matrix for it. On the contrary, [1, 5] defines a vector with two elements, and Plots has applied the same vector as the linewidth attribute of each series. So, both lines are getting a thin segment followed by a thick one. Because the number of elements in the vector given to linewidth is lower than the number of line points, Plots warns about this attribute value. The following figure shows the rendered plot:

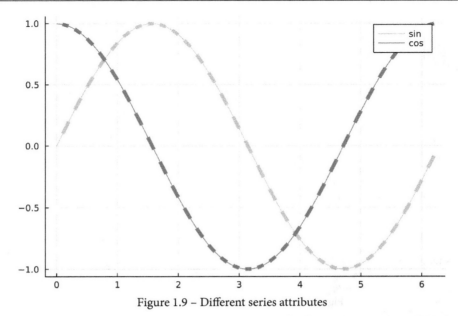

Figure 1.9 – Different series attributes

We have learned how to create multiple series in a single plot using matrix columns and a vector of vectors, ranges, or functions. While the examples only showed line plots, you can do the same for scatter and bar plots, among others. Before introducing other plots types, let's see how to add a data series to a previously created plot.

Modifying plots

Another way to add series to a plot is by modifying it using **bang functions**. In Julia, function names ending with a bang indicate that the function modifies its inputs. The Plots package defines many of those functions to allow us to modify previous plots. Plots' bang functions are identical to those without the bang, but they take the plot object to modify as the first argument. For example, let's create the same plot as *Figure 1.8* but this time using the plot! function to add a series:

1. Execute plt = plot(sin, 0, 2pi) to create the plot for the first series and store the resulting plot object in the plt variable.

2. Run plot!(plt, cos) to add a second series for the cos function to plt. This returns the modified plot, which looks identical to the one in *Figure 1.8*.

If we do not indicate the `plot` object to modify as the first argument of a `Plots` bang function, `Plots` will change the last plot created. So, the previous code should be equivalent to running `plot(sin, 0, 2pi)` and then `plot!(cos)`. However, this feature can cause problems with Pluto reactivity. So, throughout this book, we will always make explicit which plot object we want to modify.

Here, we have used the `plot!` function to add another line plot on top of a preexistent one. But the `Plots` package offers more bang functions, allowing you, for example, to add different plots types in a single figure. We will see more of these functions throughout the book. Now, let's see what other basic plot types the `Plots` package offers.

Scatter plots

We have created line plots suitable for representing the relationship between continuous variables and ordered points. However, we sometimes deal with points without a meaningful order, where scatter plots are a better option. There are two ways to create scatter plots with `Plots` – using the `plot` function and the `seriestype` attribute, or using the `scatterplot` function.

The default `seriestype` for `Plots` is `:path`, which creates the line plots. You can check that by running `default(:seriestype)`, which returns the default value of a given attribute, written as a symbol. But we can set `seriestype` to `:scatter` to create a scatter plot – for example, let's plot the `sin` function using a scatter plot:

```
plot(sin, 0, 2pi, seriestype=:scatter)
```

Most of the series types define a shorthand function with the same name and the corresponding bang function – in this case, the `scatter` and `scatter!` functions. The following code produces the same plot as the previous one, using the `seriestype` attribute:

```
scatter(sin, 0, 2pi)
```

The resulting plot is as follows:

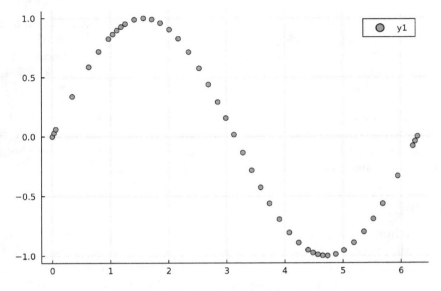

Figure 1.10 – A scatter plot

Note that the density of dots in the figure highlights the grid of *x* values that Plots created, using its adaptative algorithm to obtain a smooth line.

Bar plots

Bar plots are helpful when comparing a continuous variable, encoded as the bar height, across the different values of a discrete variable. We can construct them using the :bar series type or the bar and bar! functions. Another way to input data can come in handy when constructing bar plots – when we call the plot function using *a single vector, range, or matrix* as the first argument, Plots sets *x* to match the index number. Let's create a bar plot using this trick:

1. Run the following code:

    ```
    using Random
    heights = rand(MersenneTwister(1234), 10)
    ```

    This creates a vector of random numbers to define the bar heights. We loaded the Random standard library to make a random number generator, with 1234 as a seed to see the same plot.

2. Execute bar(heights) to create a bar plot, where the first value of heights corresponds to *x* equal to one, the second is equal to two, and so on. Note that the value of *x* indicates the midpoint of the bar. The resulting plots should look like this:

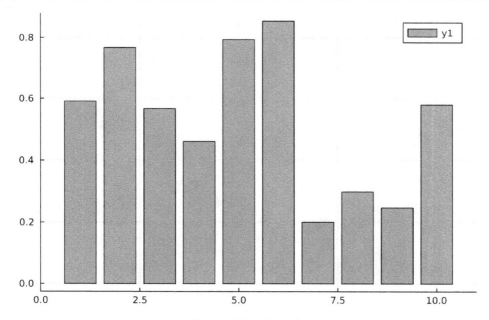

Figure 1.11 – A bar plot

You can also make *x* explicit by running `bar(1:10, heights)` on the last step; the result should be the same.

Heatmaps

The previous series type plotted a series for each column in an input matrix. Heatmaps are the plot type that we want if we prefer to see the structure of the input matrix. The magnitude of each value in the matrix is encoded using a color scale. Let's create a heatmap that matches the input matrix:

1. Execute the following code to create a 10 x 10 matrix:

    ```
    using Random
    matrix = rand(MersenneTwister(1), 10, 10)
    ```

2. Run `hm = heatmap(matrix)` to generate a heatmap. Note that the `heatmap` function plots the first matrix element at the bottom at `(1, 1)`.

3. Execute `plot!(hm, yflip=true)` to fix that. Here, the `plot!` function modifies the value of the `yflip` attribute. `yflip` puts the value 1 of the *y* axis at the top when you set it to `true`. Now, the order colors match the order of the elements in the matrix:

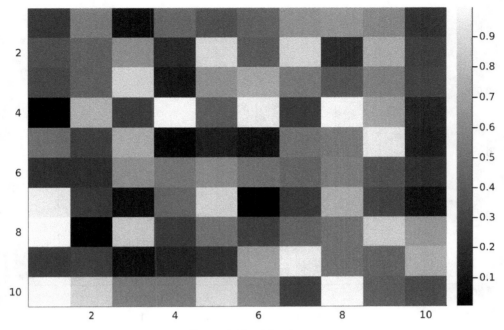

Figure 1.12 – A heatmap

We have seen how to create the most basic plot types using Plots. Let's now see how to compose them into single figures, taking advantage of the Plots layout system.

Simple layouts

Let's see the easiest way to compose multiple plots into a single figure. You can do it by simply passing plot objects to the plot function. By default, Plots will create a figure with a simple layout, where all plots have the same size. Plots orders the subplots according to their order in the attributes – for example, the following code creates a plot pane with two columns; the first column contains the plot of the sin function, and the second column the cos function plot:

```
plot(plot(cos), plot(sin))
```

In the following figure, we can see the plot created by the previous code:

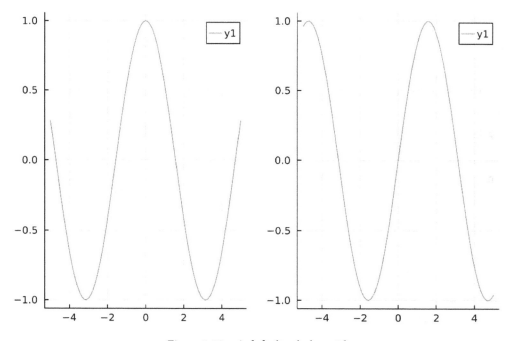

Figure 1.13 – A default subplot grid

We can use the `grid` function and the `layout` attribute of the `plot` function to customize the behavior – for example, we can have the two plots in a column rather than in a row by defining a grid with two rows and one column:

```
plot(plot(cos), plot(sin), layout = grid(2, 1))
```

The resulting plot will look like the one in the following figure:

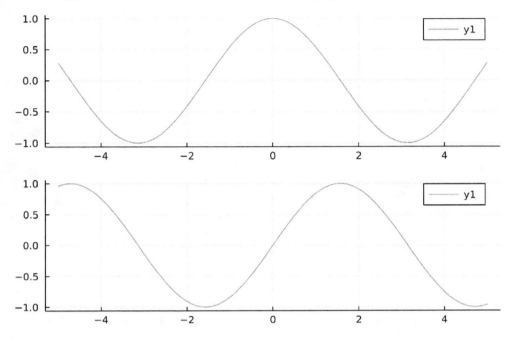

Figure 1.14 – A single-column layout

The grid function can take the widths and heights keyword arguments. Those arguments take a vector or tuple of floating-point numbers between 0 and 1, defining the relative proportion of the total width or height assigned to each subplot. Note that the length of the collection for widths should be identical to the number of columns, while the length for heights should match the number of rows in the grid layout – for example, the following code creates a panel with four plots arranged in a matrix of 2 by 2. The first column takes 80% (0.8) of the plot width, and the first row takes only 20% (0.2) of the total plot height:

```
plot(
    plot(sin), plot(cos),
    plot(asin), plot(acos),
    layout = grid(2, 2,
            heights=[0.2, 0.8],
            widths=[0.8, 0.2]),
    link = :x
    )
```

This code generates the following plots:

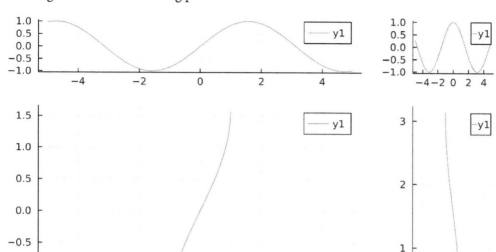

Figure 1.15 – A layout with user-defined sizes and linked x axes

Note that we have used the link attribute of a plot to link the *x* axes of each subplot column. You can use the link attribute to link the :x axes, the :y axes, or :both.

In those examples, we called the plot function inside the outer plot to create each argument, as each subplot is simple. It is better to store each subplot into variables for more complex figures. We will explore layouts in more depth in *Chapter 11*, *Defining Plot Layouts to Create Figure Panels*.

Summary

In this chapter, we have learned how to use Julia using different development environments, which will come in handy when developing interactive visualizations. We also learned how to install Julia packages, allowing us to access the Julia ecosystem's power. We also managed project environments to ensure the reproducibility of our projects through time and across computers. We have seen the most basic types and operations that you will need when creating plots in Julia. Finally, we have learned how to make some basic plots using the Plots package.

In the next chapter, we will learn more about the Julia ecosystem for data visualization. In particular, we will be going more into more depth about the `Plots` package to learn about its backends, and we will also explore Makie.

Further reading

Here, we have only scratched the surface of the Julia language. If you want to learn more about it, its documentation is the best resource: `https://docs.julialang.org/en/v1/`.

2
The Julia Plotting Ecosystem

The Julia package ecosystem offers multiple plotting options. Each of them has its strengths and weaknesses. Among those packages, two packages stand out, thanks to their fresh and innovative design: `Plots` and `Makie`. These two high-level packages allow you to choose the plotting backend that finally renders the plot.

This chapter will explore some of the most helpful packages for data visualization in the Julia ecosystem. We will explore the original philosophy behind `Plots` and `Makie` and learn about their plotting backends. Specifically, we will describe their features for allowing interactive data analysis.

At the end of this chapter, you will be able to make a well-founded decision when choosing a plotting package or backend for your data visualization task. Also, you will be able to take advantage of the backend agnostic plotting instructions that `Plots` and `Makie` offer.

In this chapter, we're going to cover the following main topics:

- Plotting libraries
- Understanding the `Plots` package
- Plots' backends
- Introducing `Makie`

Technical requirements

For this chapter, you will need Julia and IJulia installed, a web browser, and an internet connection. The IJulia notebook with the code examples and plot outputs of the different plotting libraries are in the Chapter02 folder of the book's GitHub repository: https://github.com/PacktPublishing/Interactive-Visualization-and-Plotting-with-Julia. While this chapter mentions many plotting libraries, some of them with requirements hard to install, you do not need to install them all. The only plotting libraries you need for this chapter are Plots, GLMakie, and WGLMakie.

Plotting libraries

Unlike languages such as R, Julia doesn't have a built-in plotting solution. This has motivated the Julia community to create various packages for plotting and data visualization. Some of these packages are wrappers around plotting engines from other languages, while others are pure Julia solutions. This section will briefly describe some of the multiple plotting solutions that the Julia package ecosystem offers. It will center on plotting packages with high-level interfaces that allow for data visualization. Therefore, the section will not describe packages defining plotting primitives. Also, it will not list packages derived from the listed ones.

For each package, we will show you the syntax to build a simple line plot with default attributes so that you can get a feeling for it. Also, we will show the created output for some packages. For that, we are going to use the following input data:

```
x = range(0, 2pi, length=100)
y = sin.(x)
```

You can create a temporal environment to install the plotting packages without polluting the global environment of your Julia version. To achieve that, run the following code in the Julia session in which you want to test a plotting library:

```
import Pkg
Pkg.activate(temp=true)
Pkg.add([Pkg.PackageSpec(name="ECharts", version="0.6"),
     Pkg.PackageSpec(name="GR", version="0.59"),
     Pkg.PackageSpec(name="Gadfly", version="1"),
     Pkg.PackageSpec(name="Gaston", version="1"),
     Pkg.PackageSpec(name="Gnuplot", version="1"),
     Pkg.PackageSpec(name="GracePlot", version="0.3"),
     Pkg.PackageSpec(name="InspectDR", version="0.4"),
```

```
Pkg.PackageSpec(name="PGFPlotsX", version="1"),
Pkg.PackageSpec(name="PlotlyJS", version="0.18"),
Pkg.PackageSpec(name="UnicodePlots", version="2"),
Pkg.PackageSpec(name="VegaLite", version="2"),
Pkg.PackageSpec(name="Winston", version="0.15"),
Pkg.PackageSpec(name="PyPlot", version="2")])
```

We cannot reuse the environment across Julia sessions because it is temporal. Therefore, to test the plotting packages, we need to run the previous code for each Julia session we want to use. We recommend opening a new Julia session to try each package. This is because this section's code examples use using instead of import to load the packages and avoid using fully qualified names. Therefore, as most plotting packages export the same names, such as the plot function, the ambiguities will create problems when loading different libraries in the same session. However, this is not much of a problem; we will only wait for the installation and pre-compilation of the packages the first time we execute this code.

The previous code block installs the package versions used to create this chapter's plots and code examples. This is the preferred option. However, suppose you want to install the latest version for each plotting package. In that case, you need to change the previous call to Pkg.add by adding the following:

```
Pkg.add(["ECharts", "GR", "Gadfly", "Gaston", "Gnuplot",
 "GracePlot", "InspectDR", "PGFPlotsX", "PlotlyJS",
 "UnicodePlots", "VegaLite", "Winston", "PyPlot"])
```

Note that if you have already run the first version, setting specific version numbers for each package, running the second one will not update the packages. If you want to update them, you will need to call Pkg.update() explicitly. If you find errors with the latest versions, downgrade the packages to the versions used in this book by simply running Pkg.add with the wanted versions.

It is likely that your system will not meet all the required dependencies for some of those libraries; do not worry – you can skip them. Now that we know how to set up the project environment and variables to test the packages, let's explore them.

Pure Julia packages

We are going to start by describing plotting packages written in Julia. These packages should be easy to install, as they do not require external dependencies. What's more, they enjoy the speed and composability of the Julia language. Plots and Makie, while being Julia solutions, will not be described in this section, as they are explained in depth later in this chapter.

UnicodePlots

`UnicodePlots` is a practical plotting solution that renders basic plots using Unicode characters. You can use it to inline the text-based figures in the *Julia REPL*. As it *doesn't require a graphical interface*, it is instrumental when working in environments without one. It has few features, but it supports the most common plot types: line, bar, and scatter plots. It also renders nice histograms, heatmaps, and boxplots. If your terminal's font doesn't fully support Unicode characters, `UnicodePlots` also offers `AsciiCanvas` to use ASCII characters instead.

Its syntax is simple – for example, the following code creates a line plot after executing the code indicated at the beginning of the *Plotting libraries* section of this chapter:

```
using UnicodePlots
lineplot(x, y)
```

If you execute it in the Julia REPL, you will see the plot inline, as shown in the following figure:

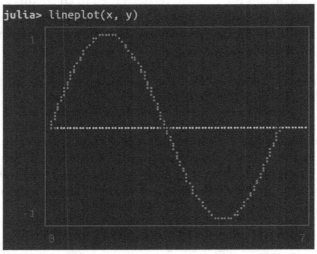

Figure 2.1 – A line plot with UnicodePlots

Next, let's explore the Gadfly package.

Gadfly

`Gadfly` is a pure Julia plotting solution relying on the `Compose` library for vector graphics. It can create complex statistical plots using simple syntax, inspired by the *grammar of graphics*. Also, it offers some basic interactivity through the use of JavaScript. `Gadfly` shines for its ability to export publication-quality figures, but it cannot create 3D plots. We will explore `Gadfly` in depth in *Chapter 5, Introducing the Grammar of Graphics.*

It has a more verbose syntax, but it is highly flexible – for example, creating a line plot will require executing the following line of code:

```
using Gadfly
plot(x=x, y=y, Geom.line)
```

That will create an interactive **Scalable Vector Graphics (SVG)** image, such as the one shown in the following figure:

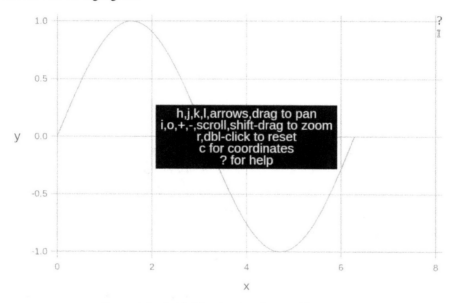

Figure 2.2 – A Gadfly plot showing the Help menu

You can see the help menu shown in the previous figure by locating the mouse in the top right corner or pressing the *?* key while the mouse is over the plot.

Winston

`Winston` is a Julia library to create 2D plots. It relies on the `Cairo` C library for rendering. The documentation isn't complete, but you will find examples and explanations in the docstring of the package functions and objects. The following code creates a line plot, as indicated by the " - " string:

```
using Winston
plot(x, y, "-")
```

We recommend opening a new Julia session and setting it up as indicated at the beginning of the *Plotting libraries* section of this chapter to execute this code. This is particularly important if you have loaded the `Gadfly` package, as both export a `plot` function. The following figure shows the line plot as it appears in the window created by `Winston` when you execute the previous code in the Julia REPL:

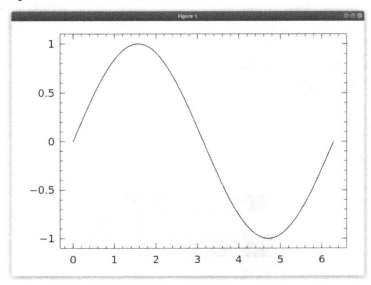

Figure 2.3 – A line plot with Winston

Let's now look at the InspectDR package.

InspectDR

`InspectDR` is another Julia package built on top of `Cairo`, therefore creating 2D figures. It focuses on rendering *fast and interactive* plots and panels. It can plot large datasets, but its syntax is complex and verbose. Let's see an example by executing the following code in a new Julia session, set up as indicated in the *Plotting libraries* section. Creating a new session is crucial if you have previously loaded `Winston`, as both packages export the `add` function:

```
using InspectDR
figure = InspectDR.Plot2D()
add(figure, collect(x), y)
display(InspectDR.GtkDisplay(), figure)
```

This code will open the **Graphical User Interface (GUI)** of InspectDR with the plot shown in the following figure:

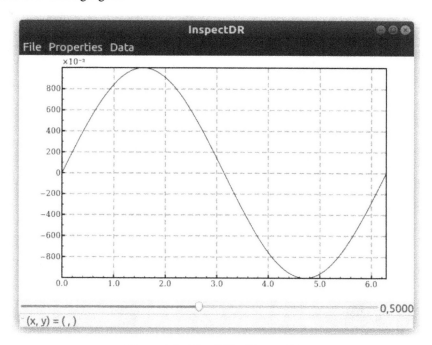

Figure 2.4 – The GUI of InspectDR

The packages mentioned offer pure Julia solutions to the plotting problem, but there are also plenty of foreign libraries solving it. Thankfully, some of those libraries have Julia wrappers. In the following section, we will explore these packages offering external plotting solutions.

Wrappers

The following packages are Julia wrappers for plotting libraries of other programming languages. Some of them take advantage of Julia's artifact system to allow for easy installation and reproducibility. Others are more cumbersome to install, as they have software requirements that your system should meet. If you already know how to plot using any of those packages in their original languages, you will quickly learn how to use its Julia interface.

PyPlot

The `PyPlot` package wraps the famous `PyPlot` module of Python's `Matplotlib`. It offers a lot of features, including 3D plots and support for LaTeX. Also, you can take advantage of the basic interactivity provided by the GUI of `Matplotlib` (see *Figure 2.5*). The following code shows a line plot using the GUI when executed in the *Julia REPL*. As before, we recommend opening a new Julia REPL. You need to set it up as indicated at the beginning of the *Plotting libraries* section. That's particularly important if you have previously loaded a library that exports the `plot` function:

```
using PyPlot
plot(x, y)
```

The following figure shows the GUI of `Matplotlib` containing the line plot:

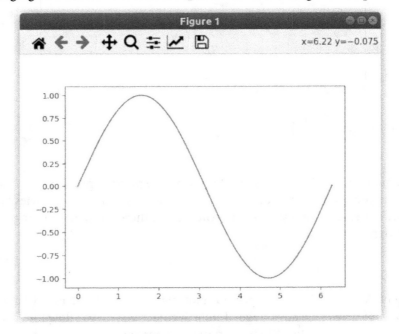

Figure 2.5 – The GUI of PyPlot

In *Pluto*, you can have the plot inline on the notebook by running the following:

```
using PyPlot
figure()
plot(x, y)
gcf()
```

PyPlot is not the only plotting library from Python wrapped in Julia. The Seaborn package offers a Julia interface to the Seaborn Python library, also based on Matplotlib.

PlotlyJS

The PlotlyJS package offers a Julia interface for the *Plotly JavaScript* library. It has many features, including tridimensional plots. Plotly, created on top of the popular *D3.js*, excels in its ability to create interactive visualizations. This package has fast and easy installation, as it automatically downloads the JavaScript library. Note that there is also the Plotly Julia package, which allows interacting with the Plotly cloud services. Execute the following code on a new *Julia REPL*, set up as indicated at the beginning of the *Plotting libraries* section:

```
using PlotlyJS
plot(scatter(x=x, y=y, mode="lines"))
```

You will see an *Electron* window, created using the Blink Julia package, with the plot, as shown in the following figure:

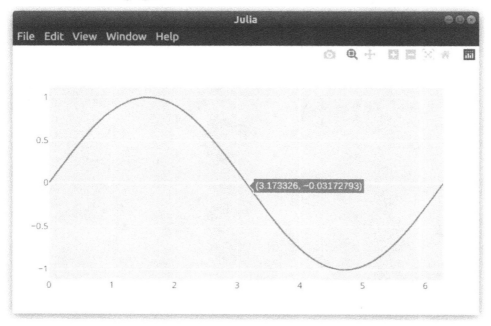

Figure 2.6 – PlotlyJS on an Electron window

You can see in the figure that Plotly shows the point coordinates on a mouse hover (the mouse pointer is not shown). Also, you can see the Plotly tools at the top-right corner of the plot. The tool icons are still available when the plot is inline on a notebook or the **Visual Studio Code (VS Code)** plot pane.

GR

GR is a Julia interface for the *GR framework*, mainly developed in C. It produces high-quality plots in a fast way. It is easy to install, as the package uses the Julia artifact system to install the GR framework automatically. Also, there is the GRUtils package that offers a more Julian interface to the GR package. You can create a simple line plot with GR by running a new Julia session after setting it up, as indicated earlier in the *Plotting libraries* section:

```
using GR
plot(x, y)
```

This opens the following window, showing the plot when you execute the code in the REPL:

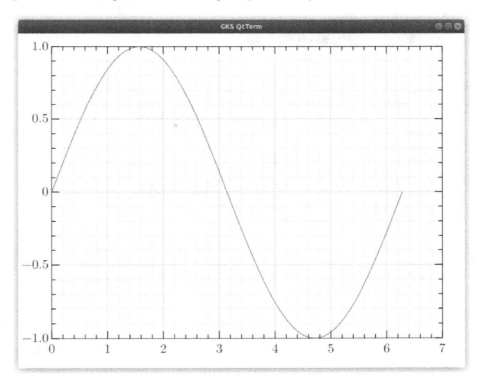

Figure 2.7 – A line plot with GR

We will look at the PGFPlotsX package next.

PGFPlotsX

PGFPlotsX wraps the *PGFPlots LaTeX package* with a similar **Application Programming Interface (API)**. PGFPlotsX can create complex figures, including 3D plots, of publication quality. Installation can be cumbersome, as it needs LaTeX and the PGFPlots package installed on your system. The line plot with this package will be as follows:

```
using PGFPlotsX
@pgf Plot({ no_marks }, Table(x, y))
```

This creates a *PDF file* containing the plot. PGFPlotsX will open it using your default application to visualize the PDF files in your operating system. The following figure shows an example of that using *Ubuntu*:

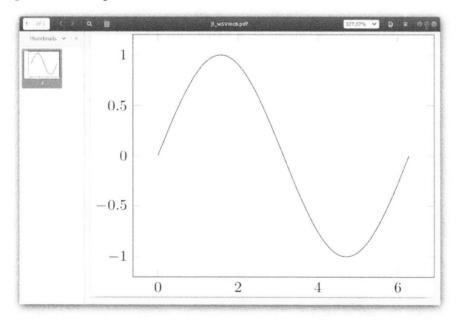

Figure 2.8 – A PDF file with the PGFPlotsX plot

Let's now look at the Gaston package

Gaston

The `Gaston` package offers a Julia interface for the classic *Gnuplot*. You should install Gnuplot on your system before using `Gaston`. It provides a lot of features, and it can produce nice tridimensional plots. Gnuplot's GUI implements some basic interactivity. Also, you can configure `Gaston` to *inline figures in the Julia REPL* using ASCII characters. You can generate a line plot example using the following code in a new Julia session, set up as suggested early in the *Plotting libraries* section:

```
using Gaston
plot(x, y, w="l")
```

This code will create the following line plot:

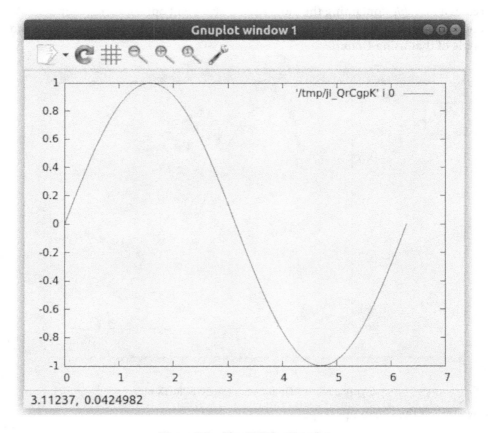

Figure 2.9 – The GUI for Gnuplot

Next, let's look at Gnuplot.

Gnuplot

As with Gaston, this also allows you to use *Gnuplot* software from Julia. But, while Gaston offers a Julian interface, this package will enable you to run Gnuplot commands directly from Julia using the @gp macro. The Gnuplot package also requires you to install the gnuplot program before it. You will notice the difference between the Gnuplot and Gaston syntaxes by comparing their code examples. The code for the Gnuplot package is as follows:

```
using Gnuplot
@gp x y "with lines"
```

The previous code opens the same Gnuplot GUI shown in *Figure 2.9*.

GracePlot

GracePlot is a Julia control interface for the *Grace* software, which offers a GUI that allows for plot customization. The interface also allows you to perform some simple interactions with the figure. You need Grace/xmgrace installed on your computer to use this package. It has a more verbose syntax. For instance, the following code creates a line plot when you execute it in a new Julia session, set up as indicated at the beginning of the *Plotting libraries* section:

```
using GracePlot
plt = GracePlot.new(fixedcanvas=false, emptyplot=false)
g = graph(plt, 0)
add(g, collect(x), y)
autofit(g)
redraw(plt)
```

The following figure shows the line plot and the GUI of Grace:

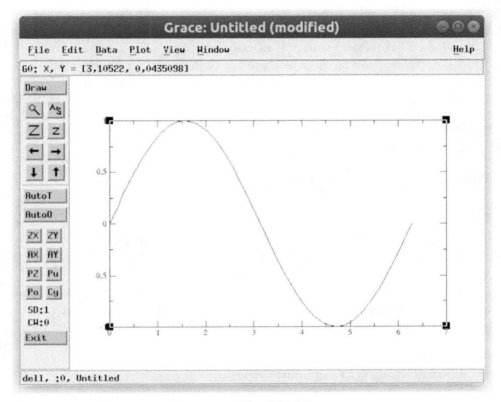

Figure 2.10 – The GUI of Grace

Let's look at the ECharts package next.

ECharts

The ECharts package offers a Julia interface for the *Apache ECharts JavaScript library*. It provides varied chart types with an easy and clean syntax. As with other JavaScript-based libraries, it excels at the creation of interactive plots. The following code will create a line plot:

```
using ECharts
line(x, y)
```

If you execute this code in the *Julia REPL*, Julia will open an *Electron* window, like the one shown in the following figure:

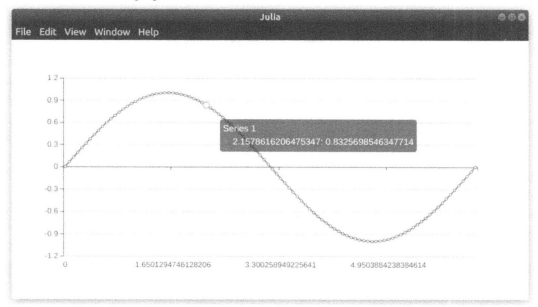

Figure 2.11 – An ECharts plot on an Electron window

The plot shows the point coordinates using the mouse hover, as shown in the previous figure (the mouse pointer is not shown).

VegaLite

The VegaLite package offers a Julian interface for the *Vega-Lite JavaScript library*, built on top of Vega. Its syntax extends the grammar of graphics into a *grammar of interactive graphics*. Indeed, the creation of interactive plots is one of its main strengths. The package offers both a high-level syntax to work with tabular data and a low-level syntax close to the one in the JavaScript package; in fact, you can declare your graphs using their JSON specifications. We will discuss this package and its syntax in more detail in *Chapter 5, Introducing the Grammar of Graphics*. A line plot using the @vlplot macro of VegaLite is as follows:

```
using VegaLite
@vlplot(:line, x=x, y=y)
```

This will open an *HTML file* in your web browser containing the VegaLite plot. The following figure shows the generated line plot and the VegaLite menu:

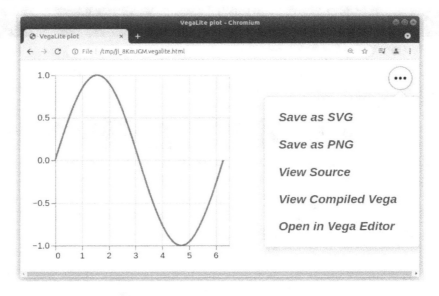

Figure 2.12 – The line plot with VegaLite

Up to this point, we have seen a lot of Julia packages for plotting. We have briefly introduced them so that you can learn their differences, strengths, and requirements. But we have skipped two of the main plotting packages in the Julia ecosystem: `Plots` and `Makie`. Now, it is time to learn about the former.

Understanding the Plots package

We saw how to create heatmaps and line, scatter, and bar plots using `Plots` at the end of *Chapter 1, An Introduction to Julia for Data Visualization and Analysis*. But we have not discussed the package, so let's do that here.

The `Plots` package is very different from the packages described in this chapter, and it has a pretty unique feature. In the previous section of this chapter, we have seen many plotting libraries available from Julia, each of them with different weaknesses and strengths. But, more importantly, those packages have different interfaces, making it difficult for a user to take advantage of all of them. That is the main problem that `Plots` solves. To achieve that, `Plots` offers a single interface that allows you to access the different plotting packages. These packages are **Plots' backends**.

When creating a plot with `Plots`, we formulate plotting instructions in a backend-independent way that the `Plots` library translates and passes to a chosen backend. The `Plots` package uses the GR package as its default backend, and it also ships with the `Plotly` *JavaScript library*. So, you can use those backends without a previous installation. If you want to use any of the others, install the backend package, as any other Julia package, before using it. We will provide a list of all of Plots' backends in the next section of this chapter.

The `Plots` interface also eases the transition from different plotting packages to `Plots` as it offers a concise and straightforward syntax, with multiple equivalent names for each attribute. For example, you can set the line width by user using either `linewidth`, `linewidths`, `lw`, `w`, or `width`. That also allows you to use contractions when quickly exploring visualizations or the most extended and descriptive attributes' names. This book will favor the latter, as they are more explicit, allowing a better understanding of the plotting instructions.

Let's see the available Plots' backends and how we can take advantage of them.

Plots' backends

The `Plots` package interfaces many of the plotting packages we described at the beginning of this chapter. Let's list Plots' backends while highlighting their strengths and weaknesses:

- `GR`: It is fast and supports most of the `Plots` features.

- `Plotly`: It creates interactive plots. It is always available.

- `PlotlyJS`: Like Plotly, but you need to install it. Also, it offers more output formats than the Plotly backend. You can update `IJulia` inline plots from any cell.

- `PyPlot`: It uses Python, which can lead to set-up and speed issues. It is a mature library that supports most of the `Plots` features.

- `PGFPlotsX`: Its dependency on LaTeX makes it hard to install, but it produces nice publication-quality plots. It supports most of the `Plots` features.

- `UnicodePlots`: It supports only a few `Plots` features. It is fast and allows for plotting in the REPL. You get better-looking bar and box plots when you use `UnicodePlots` outside `Plots`.

- InspectDR: It doesn't support many Plots features, focusing on line and scatter plots. It is fast, allowing interaction with the plot: pan, zooming, coordinate display, and adding markers to measure distances and slopes.

- Gaston: It is fast, but it misses some Plots features. You gain access to the Gnuplot GUI.

As stated in the list, not all the backends support all the features proposed by Plots.

Alternatively, the Plots series recipes can make it possible to access some backends' unsupported plot types. You can see the list of supported attributes and plot types in the Plots documentation or from a Julia session. For the latter, you need to call the Plots. supported_attrs, Plots.supported_markers, Plots.supported_scales, and Plots.supported_styles functions. Also, Plots usually warns you if you are using an attribute unsupported by the current backend, so you can change to a backend that supports it. It is nice to note that some backends allow passing some arguments directly to them using the extra_kwargs system described in *Chapter 12, Customizing Plot Attributes – Axes, Legends, and Colors*. Let's test-change the Plots backends on the Julia REPL:

1. Execute the following code in the *Julia REPL*:

```
using Plots
x = range(0, 2π, length=100);
y = sin.(x);
plt = plot(plot(x, y), scatter(x, y), bar(x, y))
```

   You will see that GR is rendering the plot. Let's check that.

2. Execute backend(). You will see that this returns the Plots.GRBackend() object, indicating that the current backend is GR.

3. Let's test Plots.supported_attrs(). You will see the list of supported attributes for the current backend.

4. Execute plotly(). This changes the current backend to Plotly. Any plot displayed now will use this backend.

5. Type plt in the Julia REPL and press *Enter*. This causes Julia to display the plot stored in the variable using the Plotly backend.

6. Test that you are now using the Plotly backend by running backend(). You will see that the function returns Plots.PlotlyBackend().

7. Execute gr() to return to the GR backend.

As you may have noticed, you can change between backends using the functions that are called as the backend name but in lowercase. Calling the backend function of an uninstalled package will give you an error, which shows the needed command to install it.

An extra backend has not been included on the previous list, as it allows you to save the plots rather than render them – the *HDF5-Plots* backend. You can choose this backend by running `hdf5()`. Then, you can use the `Plots.hdf5plot_write` function to save the plot and the data into a single file. You can read that file using `Plots.hdf5plot_read` after choosing the backend you want to use to display the stored plot. Note that it is crucial to select the backend first. Also, you can inspect the saved plot using the *HDFView* software.

We have explored how `Plots` allows us to take advantage of the multiple plotting libraries available in Julia. Now, let's explore the `Makie` package and its backends.

Introducing Makie

`Makie`, like `Plots`, is a high-level plotting package that relies on different backends for rendering. `Makie` was born from the `Plots` philosophy while trying to solve some of its problems. The strengths and issues of `Plots` come from the fact that it has high-level plotting libraries as backends. `Makie` instead relies on low-level backends, allowing it to have more control. This control permits `Makie` to be a fast plotting library with excellent interactive features.

There are three primary backends for `Makie`:

- **GLMakie**: This uses *OpenGL* to render the plots, so it is quick and allows the visualization of large datasets. `Makie` enables you to visualize the plots using a *GLFW* window. You can use it to create interactive plots, and it has excellent support for 3D plots. It is the backend that supports most of the `Makie` features. It only requires a graphic card that supports OpenGL version 3.3 or higher.

- **CairoMakie**: This focuses on the creation of publication-quality 2D plots using `Cairo`. It has limited support for 3D, and it creates static rather than interactive plots. The output plots can be inline on `IJulia` and the Pluto notebook, or in the VS Code plot panel.

- **WGLMakie**: It relies on *WebGL* to create interactive plots on the browser, which is handy when working on notebooks. It supports both 2D and 3D plots.

You should install the backend packages rather than Makie, as Makie is automatically installed and exported by these packages. Then, loading a Makie backend package is enough to access Makie. As with Plots, you have a single syntax that allows you to create plots using different backends, and you can change the backend during the sessions. For example, you can use GLMakie or WGLMakie to interactively explore your data and move to CairoMakie to create the final figure.

Let's create some basic plots using Makie to get a feeling for it:

1. Install the GLMakie and the WGLMakie packages in a temporal project environment by running the following code in the *Julia REPL*:

```
import Pkg
Pkg.activate(temp=true)
Pkg.add([
Pkg.PackageSpec(name="GLMakie", version="0.4"),
Pkg.PackageSpec(name="WGLMakie", version="0.4")
])
```

2. Execute using GLMakie to load Makie and the GLMakie backend.

3. Execute the following code to generate the example data:

```
x = range(0, 2π, length=100)
y = sin.(x)
mat = [sin(i) * sin(j) for i in x, j in x]
```

4. Execute lines(x, y) to create the most simple *line plot*. You will see the interactive OpenGL-generated plot in the GLFW window (see *Figure 2.13*). The first plot can take some seconds to be displayed; the following ones will be faster.

5. Execute scatter(x, y) to create a *scatter plot* with the same data on the same window.

6. Execute barplot(x, y) to create a simple *bar plot*.

7. Run plt = heatmap(mat) to create a simple *heatmap* and to store the plot object.

8. Execute using WGLMakie to load the WGLMakie backend.

9. Run WGLMakie.activate!() to change the current backend to WGLMalkie.

10. Type plt in the Julia REPL and press *Enter* to render the heatmap using the current backend. You will see that WGLMakie is using your browser to show you the heatmap plot.

The following figure shows the GLFW window with the line plot generated in *step 4* using GLMakie:

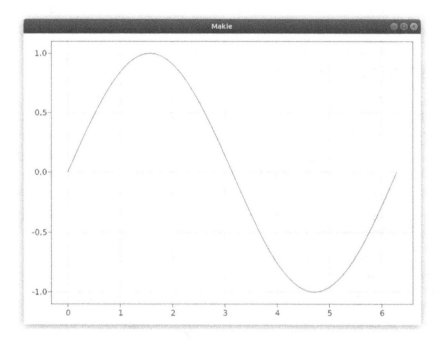

Figure 2.13 – A line plot in the GLMakie window

This has been a quick and gentle introduction to Makie, where you have learned how to create simple plots and move between backends. We are going to further explore this plotting library and its interactivity throughout this book.

Summary

In this chapter, we have learned about the multiple plotting libraries available in the Julia ecosystem, and we paid particular attention to the Plots and Makie libraries. We have learned how those libraries allow us to create a single plot that you can later render using different backends. This versatility is a powerful feature of Plots and Makie, and you will be able to make the most out of it. Finally, we have introduced how to make some simple plots using Makie. Throughout this book, we will learn more in depth about Plots and Makie and other libraries we have seen in this chapter, such as Gadfly and VegaLite.

In the next chapter, we will learn about the interactivity offered by some of the packages we have explored in this one. Furthermore, we will learn the multiple tools that Julia offers to create interactive visualizations and how to take advantage of those tools using `Plots` and `Makie`.

Further reading

- We recommend taking a look at the `Plots` and `Makie` documentation. In particular, you will learn more about Plots' backends here: `http://docs.juliaplots.org/latest/backends/`.

- You can look at the *Examples* section to see how different backends render the same plots. Finally, there is a table with the supported attributes for each Plots' backend here: `http://docs.juliaplots.org/latest/generated/supported/`.

- Information about the `Makie` backends is here: `https://makie.juliaplots.org/stable/documentation/backends/`.

- We also recommend taking a look at Makie's publication: *Makie.jl: Flexible high-performance data visualization for Julia, The Journal of Open Source Software, Simon Danisch and Julius Krumbiegel, (2021), 6(65), 3349* (`https://doi.org/10.21105/joss.03349`).

3

Getting Interactive Plots with Julia

Interactive visualization is critical for analyzing large and complex datasets, where static plots can show only some aspects. Through interaction, it is possible to explore data more profoundly, which could lead us to formulate new hypotheses or gain further insight into the task at hand.

Julia is a dynamic programming language that lets us be in the loop thanks to its advanced **read-eval-print loop** (**REPL**). However, the interactivity we can get with Julia goes even further, thanks to a whole series of packages that significantly improve our experience when analyzing and visualizing data. We have introduced some of those packages in the previous chapters: `Plots`, `Makie`, `Pluto`, and `IJulia`. In this chapter, we will learn how to use them to create interactive visualizations.

You will learn in this chapter many ways to get interactive visualizations using Julia. We will explore interactive plotting libraries, emphasizing `Makie` and `Plotly` as a backend of `Plots`. We will also learn how to use the `Interact` and `PlutoUI` libraries to create interactive visualizations using `Jupyter` and `Pluto` notebooks, respectively. The chapter will also explain Julia's tools to publish and share interactive visualizations using dashboards, web pages, and simple **user interfaces** (**UIs**).

In this chapter, we're going to cover the following main topics:

- Libraries focusing on interactivity
- Exploiting the interactivity of `Plots` backends
- Interactive and reactive plots with `Makie`
- Interactivity on `Jupyter` and `Pluto` notebooks
- Sharing your interactive visualizations

Technical requirements

You will need a computer with an internet connection and the following software installed:

- A modern web browser.
- Julia 1.6 or higher.
- The `Pluto` and `IJulia` Julia packages to access the notebooks and `Plots`, `GLMakie`, and `WGLMakie` for plotting. The chapter will also mention other plotting libraries, but those are the only ones that are required.
- A text editor could be handy to work with the example scripts; we recommend having **Visual Studio Code** (**VS Code**) with the *Julia extension.*

The code examples for this chapter are in the `Chapter03` folder of the book's GitHub repository, found at `https://github.com/PacktPublishing/Interactive-Visualization-and-Plotting-with-Julia`.

Libraries focusing on interactivity

Different plotting libraries offer different degrees of interactivity out of the box. Also, there are packages we can use to gain interactivity, even using static plotting libraries. Here, we will classify and describe the available actions while mentioning the Julia packages that provide them. In particular, we will analyze two kinds of interactions: the ones affecting the underlying data and the ones acting at the perceptual level. Let's start with the former.

Modifying the underlying data

As Julia is a dynamic language with a flexible REPL, you can always modify data and redraw the plots, but there is a series of tools that make that task easier. For example, the reactivity of `Pluto` allows us to modify the data in one cell and automatically get the related plot updated on another. `Makie` also offers a way to react to changes in variables through the use of the `Observables` package. The `Interact` package, built around `Observables`, also allows the modification of input parameters through UI elements. We will describe the interaction through `Pluto`, `Makie`, and `Interact` in more depth in future sections of this book. So, let's first examine the `Observables` library.

Triggering actions using Observables

The `Observables` package exports the `Observable` type, which wraps a Julia object. We can define functions that would be triggered when Observable's value changes. First, let's explore creating an observable that prints its value every time it changes. Proceed as follows:

1. Type `julia` on a terminal and press *Enter* to access the *Julia REPL*.

2. Press the] key, type `add  Observables`, and press *Enter* to install the `Observables` package.

3. Press *Backspace* to return to the *Julia prompt* and execute `using  Observables` to load the package.

4. Copy the following code and paste it into the Julia REPL:

    ```julia
    julia> name = Observable("World")
    Observable{String} with 0 listeners. Value:
    "World"
    ```

    As you will see, `Observable("World")` is creating an `Observable` object containing the value `"World"`. `Observable` is a parametric type; the `Observable` object we instantiated in this example can only store values of type `String`. Note also that it doesn't have listeners; let's assign one to it.

5. Execute the following code:

    ```julia
    julia> name _ observer = on(name) do value
               println("Hello $value")
           end
    (::Observables.ObserverFunction) (generic function with 0
    methods)
    ```

The on function takes a listener function as the first argument; we create one using the do syntax. The second argument of the on function is the `Observable` object we want to observe. The code returns an `ObserverFunction` that we can use to later remove the listener from the observable using the `off` function.

6. Type name and press *Enter*; you will see that name now has one listener.

7. Execute name[]; this will return the current value wrapped in the Observable object.

8. Run name[] = "Julia"; You will see that a **"Hello Julia"** message is printed on the screen, as the change on the observable value triggers the execution of the listener function. Note that we also needed the two *square brackets* to modify the wrapped value. Forgetting the square brackets will cause the Observable object to be lost.

9. Type name and press *Enter* to see the Observable object; you will notice that it now has one listener and that the current value is **"Julia"**. So, let's test how to remove the listener.

10. Execute off(name_observer) to remove the listener we assigned in *Step 5*.

11. Type name and press *Enter*; notice that the Observable object has no listeners now.

12. Execute name[] = "World" to change the observable value; notice that Julia doesn't print the welcome message because we removed the listener.

By following the previous steps, we now have a good feeling about working with Observables in Julia. Now, we will learn how to use Observables to get interactive visualizations using Plots with the GR backend on the Julia REPL. Let's follow the next steps to create an arc that updates every time we change its angle:

1. Type julia on a terminal and press *Enter* to access the Julia REPL.

2. Execute using Observables, Plots to load both packages.

3. Run angle = Observable(pi/2) to create an observable that can take float64 values containing a pi/2 value. We will use it to update the central angle that defines the arc of the plot.

4. Execute the following code:

```
function plot_arc(angle_value)
    angles = range(0, angle_value, length=100)
    x = cos.(angles)
    y = sin.(angles)
    plt = plot(
        x, y,
        ratio=:equal,
        xlims=(-1.5, 1.5), ylims=(-1.5, 1.5),
        legend=:none,
```

```
        framestyle=:none)
    display(plt)
end
```

The previously executed code defines a `plot_arc` function that takes an angle value and displays the plot showing the arc for that angle. We have set the `legend` and `framestyle` plot attributes to `:none` to hide all decorations and let only the arc display. We have also assigned the *x* and *y* axes limits so that the arc doesn't move when the angle changes. We must call the `display` function on the plot object to ensure that the plot is redrawn every time this function is triggered.

5. Execute `angle_observer = on(plot_arc, angle)` to assign the previous function as a listener to the `angle` observable. There is no plot yet, as there are no changes on the `angle` variable triggering the listener.

6. Execute the following line of code:

    ```
    notify!(angle)
    ```

    This triggers the listener function using the current value of the `angle` observable. The `notify!` function will be helpful in a situation when you need to trigger the listeners manually because the change is not automatically detected—for example, when pushing values into an observable containing a collection.

7. Execute `angle[] = pi*2/3;` to enlarge the arc. You will notice that Julia has updated the plot on the same window.

You can now change the value of the angle, and Julia will update the plot accordingly. But be careful, as this has worked like a charm on the *Julia REPL* using the *GR backend* of `Plots`. However, a different backend could create a separate window for each new plot instead of redrawing it on the same window. That problem also happens on VS Code, even with the GR backend. This example also works on `Jupyter` using the `IJulia` package, but it will display the plot under each cell that triggers the listener. Also, `Jupyter` will display multiple plots if the listener function is activated many times in a single cell. Later in this chapter, we will see better ways to create interactive plots in `Jupyter`. Also, we will explore `Makie`, a library that, contrary to `Plots`, was designed from the ground up to work seamlessly with `Observables`.

Selecting data

We have mentioned ways to modify data on the Julia side and use the modified data to re-render the plot. We can take advantage of that to filter, select, and subset data. However, the methods mentioned previously don't allow us to perform those actions by interacting with the plot object. Luckily, the `Immerse` package offers a **graphical UI (GUI)** for

Gadfly, extending the interactivity of the library and offering *point and lasso selection* out of the box. Once you have selected points using the lasso selection tool, you can determine the Julia variable to store the values. Also, Immerse allows us to add specific behaviors to point selections using the hit function—for example, you can use this to keep the coordinates of the selected point on a Julia variable. Also, the setproperty! function of Immerse can be used to modify the plot object from the Julia REPL interactively. Note that you can get similar functionally with Makie using the GLMakie backend, as we will see later.

Modifying data's perceptual formatting

Here, we will discuss the most popular actions on interactive UIs for data visualization. These actions act at the perceptual level, changing the visual aspects of a plot. Let's see these actions while listing the packages that offer them, as follows:

- **Pan**: Panning is a standard action that allows you to navigate through a plot. Multiple packages allow it out of the box: Gadfly, InspectDR, PyPlot, Plotly/PlotlyJS, Gaston, Gnuplot, GracePlot, GLMakie, and WGLMakie. When using ECharts, you can pan on the *x* axis by calling the slider! function on the plot object. You can also allow panning on VegaLite by setting the selection keyword argument of the @vlplot macro to {grid={type=:interval, bind=:scales}}.

- **Zoom**: Zooming is another classical action to navigate plots. Many Julia packages support it: Gadfly, InspectDR, PyPlot, Plotly/PlotlyJS, Gaston, Gnuplot, GracePlot, GLMakie, and WGLMakie. You can also zoom on ECharts plots by calling the slider! function on the plot object. For VegaLite, you need to set the selection keyword argument of @vlplot to the value indicated for panning.

- **Viewpoint change**: When exploring a tridimensional plot, it is common to change the viewpoint by rotating it. The PyPlot, Plotly/PlotlyJS, Gaston, Gnuplot, GLMakie, and WGLMakie libraries allow this action.

- **Display coordinates**: The following plotting libraries show the coordinate location of the mouse, sometimes through tooltips: Gadfly, InspectDR, PyPlot, Plotly/PlotlyJS, Gaston, Gnuplot, GracePlot, and ECharts. Plotly, PlotlyJS, and ECharts stand out on that list, as they show the actual data point coordinates rather than the general coordinates of the image. When using GLMakie, you need to call DataInspector on the figure, axes, or scene object to gain this feature.

- **Labeling**: Sometimes, it is helpful to add marks to data points or see their labels when interacting with them. Plotly and PlotlyJS excel on that last action, as you can personalize the point labels that appear on the tooltips. When using Plotly or PlotlyJS as a backed for Plots, you can use the hover keyword argument to personalize the tooltip text. InspectDR, Gnuplot, and Gaston allow the marking of points on the figure. Using InspectDR, you can label a reference point using the R key. Then, you can select multiple points by pressing the D key. That adds a label showing the difference in the x and y coordinates and the corresponding slope. The Gaston and Gnuplot GUIs offer similar functionality, as you can mark a point using the R key to measure the distance between that reference point and the mouse pointer for the x and y axes.

- **Show/hide**: Hiding or showing particular series can help us to understand complex datasets better. Plotly, PlotlyJS, Gadfly, and ECharts offer this functionality out of the box, whereby you can hide or show a specific group by clicking it on the color legend.

- **Layout**: The GUI of some packages allows you to perform changes on the plot layout. In particular, InspectDR allows you to adjust plot sizes, and PyPlot allows you to modify the plot's margins and spacing.

Some plotting libraries have full-featured GUIs, allowing more actions to modify the figure's visuals. In particular, the GUI of PyPlot allows you to change the title and the axes' labels, limits, and scales. It also allows you to change the style of lines and markers. However, those features cannot work as expected when using PyPlot as a backend of Plots. The GUI of GracePlot also offers a similar capability to change plot axes, titles, colors, and line and marker styles.

Note that the Julia packages wrapping JavaScript libraries allow you to customize their interactive behaviors, allowing actions that are not present out of the box. Those packages are Plotly, PlotlyJS, ECharts, and VegaLite. You can look at the documentation of the underlying JavaScript libraries to find out more about the possibilities. As with JavaScript, the GLMakie backend of Makie allows you to customize a plot based on mouse and keyboard events, therefore giving a lot of options for interactions. The InteractiveViz package uses this GLMakie feature to enable zooming and panning massive datasets, as it calculates the needed points on demand. As we will see later in this chapter, GLMakie also offers a set of interactive widgets you can use to create your interactive visualizations.

Exploiting the interactivity of Plots backends

In the previous section, we have seen that multiple `Plots` backends offer interactivity out of the box—in particular, `Gaston`, `InspectDR`, `PyPlot`, `Plotly`, and `PlotlyJS`. Among those, `Plotly` and `PlotlyJS` provide the largest number of available interactive actions. So, let's try the interactive capabilities of `Plots` utilizing the `Plotly` *backend*, as follows:

1. Open the Julia REPL and execute `using Plots` to load the `Plots` package.

2. Execute `plotly()` on the Julia REPL to use the `Plotly` backend that comes with `Plots`.

3. Run the following code to generate example data:

    ```
    x = 0:0.1:2pi
    y_sin = sin.(x)
    y_cos = cos.(x)
    ```

4. Execute `plot(x, [y_sin y_cos], labels=["sin" "cos"])` to create a plot that would be rendered with `Plotly`. Let's interact with it.

5. Move the mouse over the plot; you will see that a *toolbox* appears on the top right. The options on the toolbox depend on the plot type, and they can also change between the `Plotly` and the `PlotlyJS` backends or if we use it from VS Code. We can see that the **Zoom** button is activated (see *Figure 3.1*), so let's zoom in to our plot.

6. Press the *left mouse button* on the plot area and move the mouse while pressing that button. If you move the mouse horizontally, you will zoom in on a section on the *x* axis. Move the mouse vertically to zoom in on *y* or go diagonal to zoom in to the selected box. Release the left mouse button to zoom. Now we have zoomed in to a plot region, we will use panning to explore the plot.

7. Go to the *toolbox* on the top right and click the **Pan** button. You will see that the mouse pointer changes from a cross to a hand when you move the mouse over the points.

8. Press and hold the *left mouse button* on the plot area; you can now move the mouse to explore different plot areas while keeping the zoom level.

9. Go to the *toolbox* and click on the **Zoom out** button (the one with the minus symbol) to zoom out of the plot while keeping the current place.

10. Click the **Reset axes** button (the house symbol) to come back to the original plot axes.

11. Go to the *plot legend* and click the **cos** label. You will see that the series label gets more transparent, and the series disappears from the plot to focus on the other ones.

12. Click on the **cos** label again to show that series again.

13. Move the mouse over the lines; you will see that a *tooltip* appears showing the coordinates of the point and the series label. Let's customize the information we get on the tooltip.

14. Execute the following code:

```
_ round(number)  = round(number, sigdigits=4)
point _ label(x, y, fun)  = "$fun($( _ round(x))) = $( _
round(y))"
tooltips _ sin = [point _ label(x, y, "sin") for (x, y) in
zip(x, y _ sin)]
tooltips _ cos = [point _ label(x, y, "cos") for (x, y) in
zip(x, y _ cos)]
```

This code has created two variables: `tooltips_sin` and `tooltips_cos`. Each of them holds a vector of strings. Those strings are the labels or tooltips we want to show for each data point. To create vectors, we have first defined two helper functions: `_round` and `point_label`. The former is simply a wrapper for the `round` Julia function to round a floating-point number to four significant figures. The latter takes two numbers, the *x* and *y* coordinates, and a function and returns a string. It uses string interpolation to create an output string from a set of variables and expressions. Julia **string interpolation** uses the $ character inside a string literal to interpolate the value of the variable or expression appearing in the string. For that, you can write the variable names just after the $ character, but you should surround complex expressions with parentheses. In this case, we have interpolated the function name and the rounded x and y values.

Finally, we created a vector of strings needed to define hover text for each series using array comprehension. In the *Collections* section of *Chapter 1, An Introduction to Julia for Data Visualization and Analysis*, you can read more about array comprehension. Here, instead of simply iterating over a single collection, we use the `zip` function to iterate two collections simultaneously. In this example, the `zip` function returns a tuple of x and y values in each iteration. We are performing a **destructuring assignment** of the tuple, assigning its first element to x and the second to y when we write the `(x, y)` tuple in the `for` loop of the array comprehension.

15. Run the following code line:

```
plot(x, [y _ sin y _ cos], labels=["sin" "cos"],
hover=[tooltips _ sin tooltips _ cos])
```

Here, we plot the same data that we plotted in *Step 4*, but we define the text we want to display on mouse hover. We achieve that by using the hover plot attribute. Because we have two series, we passed a matrix with one row and two columns. Each column contains a vector of labels that we want to assign to each point. In Julia, an expression such as [x, y], where commas separate the elements, defines a vector. In contrast, [x y], which uses spaces rather than commas to separate the components, represents a matrix.

16. Move the mouse over the lines; you will see that the tooltip now displays the text we have assigned for each point of the series.

We can see the output of this process and the customized tooltips in the following screenshot:

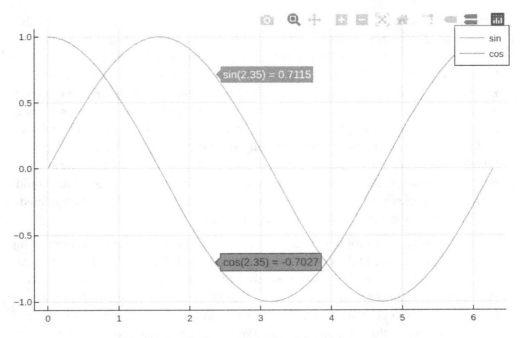

Figure 3.1 – Plotly-customized tooltips on mouse hover

Great! We have seen how to make the most of the interactivity that the Plotly and PlotlyJS backends of Plots offer. Note that if we move across the plot through panning and zooming, it doesn't change on the Julia side. So, to save the current view, we need to save the plot using the **Download plot as png** button on the `Plotly` toolbox or the **Save plot** button on the VS Code plot pane. We face a similar problem with `PyPlot`, `Gaston`, or `InspectDR`; if you change your plot aspect using their GUIs, you need to save the plot using the **Save** or **Export** button of `PyPlot` and `Gaston` or the **Export** option of the **File** menu on `InspectDR`.

In these sections, we have discussed different interactive actions we can take with various plotting libraries, focusing on the backends of `Plots`. Now, let's focus on the customizable interactivity of `Makie`.

Interactive and reactive plots with Makie

`Makie`—in particular, its `GLMakie` backend—was designed to be interactive. `Makie` builds its interactivity around the `Observables` package. For example, `Makie` arguments and attributes can be observables, so the plot automatically reacts to their changes. What is more, `Makie` extends `Observables` to make actions helpful in creating more straightforward interactive visualizations. In this section, through an example, we will learn how to create a simple interactive plot while exploiting the tools offered by `Makie`. We will plot an arc with `GLMakie` using a slider to choose the internal angle value. Proceed as follows:

1. Open the *Julia REPL* on a terminal and execute `using GLMakie`, to use `Makie` and its **Open Graphics Library** (**OpenGL**) backend. Depending on your machine or the need for precompilation, this step and the following two can take some time.

2. Execute `fig = Figure()` to create a new figure. The `Figure` objects of `Makie` contain all the information of the plot's layout, together with the objects on it. For the moment, the new figure is empty, but we will add and arrange objects on it in the following steps. *Chapter 10, The Anatomy of a Plot*, will describe in depth the plot components of `Makie`.

3. Execute the following code:

```
angle_selector = Slider(
fig[2, 1],
range = 0:0.01:2pi,
startvalue = pi/2)
```

This code creates a `Slider` object located in the second row and first column of the figure's layout. We indicated this when we indexed the `Figure` object using `[2, 1]`. The `Slider` type defines a `Layoutable` object from the `MakieLayout` module of `Makie`. Layoutable objects have their position and size controlled by `GridLayout` of a `Figure` or `Scene` object. We will see more of the advanced layout capabilities of `Makie` in *Chapter 11, Defining Plot Layouts to Create Figure Panels.*

In this case, we have used the `Slider Layoutable` object to add a slider widget to our figure. The slider will allow us to choose a value between `0` and `2 pi`, with a step size of 0.01, as indicated by the `range` keyword argument. Then, we have set the `startvalue` keyword argument to `pi/2` to define a starting value for the interactive slider. Let's explore the created `Slider` object before continuing.

4. Type `angle_selector.value` to see the value stored in the slider. Note that the `value` field contains an `Observable` object. This `Observable` object works as described in the *Triggering actions using Observables* section of this chapter.

5. Go to the `Makie` window and change the value of the slider. You can do that by moving the slider's thumb along the track while pressing the left mouse button.

6. Execute `angle_selector.value` to see that the stored value has changed. As the slider changes the value of an `Observable` object, we can define listeners for it.

7. Execute the following code:

```
angles = lift(angle -> range(0, angle, length=100),
angle _ selector.value)
```

The `lift` function of `Makie` returns an `Observable` object that listens to other observables. It takes at least two positional arguments; the first should be a function, and the rest are the listened observables. The passed function will determine the value of the returned observable, and it should take as input the values of the listened observables.

In this example, we listen to the `Observable` object stored in the `value` field of the `angle_selector Slider` object. The new `angles Observable` object will determine its value by applying the anonymous function we gave to the `lift` function. Therefore, `angles` will contain a range of 100 values, from 0 to the number we have selected with the slider.

8. Run the following line of code:

```
points = @lift([Point2(cos(angle), sin(angle))   for angle in
$angles])
```

In this case, we used the @lift *macro* equivalent to the lift function but it is less verbose. With the @lift macro, we only need to write the body of the function we want to apply to the listened observables. In that expression, we need to add a $ prefix to the Observable variables whose changes will trigger the change of the new one. The object returned by the given expression will determine the value contained in the new observable.

In our example, the expression is simply an *array comprehension*. We are listening to changes in the angles Observable object, whose modification will modify the values in the new points Observable object. The new observable will contain a vector of the arc's points we need to draw given a set of angles. We are using the Point2 type of Makie to define bidimensional points.

9. Execute the following code:

```
lines(fig[1,1], points,
      axis=(aspect=DataAspect(),
            limits=(-1.5, 1.5, -1.5, 1.5)))
```

Here, we have used the lines function to create a line plot that follows the coordinates stored in the points observable. We have located the line plot in the first row and column of the figure's layout, as indicated by fig[1, 1]. In that code, we also set the axis keyword argument to fix the aspect ratio and the axes limits to see the unit circle with its correct proportions. You can learn about these axes customizations in *Chapter 12, Customizing Plot Attributes – Axes, Legends, and Colors*.

10. Move the slider to see how the plot changes. Every change on the slider triggers the calculation of the angles observable, which triggers the computation of the points observable. As the lines function takes the points Observable object as an argument, the plot gets updated when it changes.

Great! We have created our first interactive Makie plot, and we have taken advantage of the Slider widget. Makie also offers other widgets—in particular, IntervalSlider, Menu, Textbox, Toggle, and Button. The slider we created in the previous plot has no labels. We can create custom labels using the Label object, but luckily, Makie offers the labelslider! Utility function to create a slider with a name and a label that shows its current value. Let's execute the following code to get such a slider for our arc:

```
using Printf
fig = Figure()
_ angle _ label(angle) = @sprintf "%3.2fπ" angle/pi
```

```
angle_selector = labelslider!(fig, "angle", 0:0.01:2pi, format=_
angle_label)
set_close_to!(angle_selector.slider, pi/2)
fig[2, 1] = angle_selector.layout
angles = lift(angle -> range(0, angle, length=100), angle_
selector.slider.value)
points = @lift([Point2(cos(angle), sin(angle))  for angle in
$angles])
lines(fig[1,1], points, axis=(aspect=DataAspect(), limits=(-1.5,
1.5, -1.5, 1.5)))
```

We can see a few changes in this last code block compared to the one we achieved in *Step 9* of the previous instructions in this section. First, we loaded the Printf standard library to use its @sprintf macro to interpolate a value into a string following a specific format. In this case, we define an _angle_label function that will format the slider's current value using @sprintf to ensure that all the values are the same width—a floating-point number of at least three digits with two decimal places.

Then, we created a slider using the labelslider! function, which takes the parent figure, the label for the slider, a range of values as positional arguments, and the function used to format the numbers using the format keyword argument. It returns a named tuple containing the slider and the layout, among other things. We access the slider and set its starting value to be close to pi/2 using the set_close_to! function.

Finally, we assign the slider layout to the second row of the first column of the figure. Note that the lift function needs to access the value observable of the slider stored in the tuple named angle_selector. The rest of the code remains the same. If you want to control more variables using multiple sliders, you can take advantage of the labelslidergrid! function.

The previous code block using the labelslider! function will create a plot that has a labeled slider, as shown in the following screenshot:

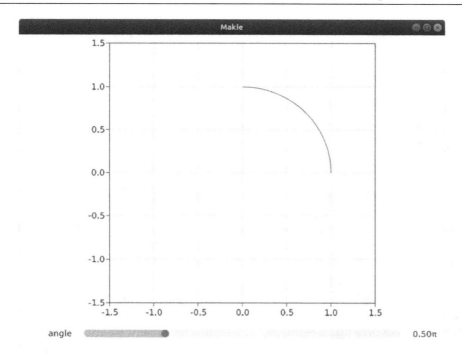

Figure 3.2 – Labeled slider of GLMakie

Besides the widget utilities and the observables as inputs, `Makie` also returns *mouse and keyboard interactions* as objects of type `PriorityObservable`. The `PriorityObservable` Makie type acts as an `Observable`, but it also allows the setting of a priority level to the listeners so that a listener function with high priority can stop the event's propagation by calling `Consume(true)`. You can access the mouse and keyword interaction by calling the `events` function on the figure's and axes' `scene` attribute. That returns an object of type `Events`, containing a `PriorityObservable` object associated with different mouse and keyboard events.

To make it easy to handle these interactions, `Makie` also offers a series of convenient functions such as `mouseposition` that return the mouse's position in data coordinates. You can use those elements to customize the interaction with the plot entirely. Let's learn how to deal with mouse events by determining the angle of our arc using the mouse position rather than the slider by running the following code:

```
fig = Figure()
axis = Axis(fig[1,1], aspect=DataAspect(), limits=(-1.5, 1.5, -1.5,
1.5))
angle = Observable(pi/2)
points = @lift([Point2(cos(a), sin(a)) for a in range(0, $angle,
```

```
        length=100)])
    lines!(points)

    fig_events = events(axis.scene)
    on(fig_events.mouseposition) do event
        x, y = mouseposition(axis.scene)
        angle[] = y > 0 ? atan(y, x) : 2pi + atan(y, x)
    end
```

Here, we first created an empty figure calling the Figure type constructor. Then, we made axes for the figure through the Axis type constructor, and we located those axes in the first cell of the grid layout. We will see in more depth the different parts of a figure and how to organize its layout in *Chapter 10*, *The Anatomy of a Plot*, and *Chapter 11*, *Defining Plot Layouts to Create Figure Panels*, respectively. We store the Axis objects to access its scene field later. That is needed to get the *mouse positions* relative to the axes' coordinates, as accessing the figure's scene will give us figure coordinates in pixels. Then, we created an angle Observable object and a points Observable object that listens to the changes on angle using the @lift macro. Finally, we used the lines! function to add a line following the coordinates stored in points to the current figure.

After executing the lines! function, we have a line plot that can react to changes in angle but that doesn't listen to changes in the mouse position. To achieve the last point, we need to obtain the Events object of the scene field of Axis by using the events function. Then, we add a listener to the mouseposition PriorityObservable object wrapped in the Events object with the on function. As described in the *Triggering actions using Observables* section of this chapter, the on function takes a listener function that we can create with the do syntax. In this example, the listener uses the mouseposition function on the axes' scene field to get the position coordinates of the mouse. Note that we use *destructuring assignment* to assign the first and second values of the Point object returned by mouseposition into the x and y variables, respectively. We update the value of the angle observable using the x and y mouse coordinates, therefore triggering the change on the plot.

As you can see, Makie offers excellent interactivity out of the box and allows you to fully customize the interaction with your visualization. We have learned how to take advantage of that by using widgets and assigning behaviors to mouse or keyboard events. Now, let's see how we can use notebook widgets to create interactive visualizations, even through static plotting libraries.

Interactivity on Jupyter and Pluto notebooks

As `Pluto` and `Jupyter` notebooks run in the browser, we can use web technologies—namely, **JavaScript**, **HyperText Markup Language** (**HTML**), and **Cascading Style Sheets** (**CSS**)—to interact with our visualizations. This section will showcase some high-level tools, mainly `PlutoUI` and `Interact`, which will help us use those technologies' power from Julia while using `Pluto` and `Jupyter` notebooks. Julia plotting libraries that wrap JavaScript ones are suitable for creating interactive plots within notebooks. However, we will see tools that allow us to interact with the visualization, even with static plotting libraries. Let's dive first into the interactive features of `Pluto` for data visualization.

Creating interactive visualizations with Pluto

We explored the reactive behavior of `Pluto` in *Chapter 1*, *An Introduction to Julia for Data Visualization and Analysis*. Each time you update a cell, all the cells that depend on that one get automatically re-executed. That makes it easy to get the interactive behavior described in this chapter's *Triggering actions using Observables* section without needing `Observables`. Let's explore that by recreating the same arc plot of that section using `Pluto`. The plot will react to changes in the underlying data, being re-rendered after any change on the internal angle value. Proceed as follows:

1. Open a Julia terminal and execute `import Pluto; Pluto.run()` to open `Pluto` in a new browser tab.

2. Click on the **New Notebook** link, which will create a new empty `Pluto` notebook.

3. Type `using Plots` in the first cell and press the *Ctrl* and *Enter* keys together to run that cell while adding a new cell below. `Pluto` is going to install `Plots` automatically on the notebook's project environment. For this example, as with the previous one, we will use the default *GR backend* of `Plots`.

4. Execute `angle = pi/2` in the new cell and create a new cell below.

5. Run `angles = range(0, angle, length=100)` in the empty cell.

6. Run the following code in a new cell:

```
plot(cos.(angles), sin.(angles),
    ratio=:equal,
    xlims=(-1.5, 1.5), ylims=(-1.5, 1.5),
    legend=:none,
    framestyle=:none)
```

You will see the static arc plot is automatically inline over the cell.

7. Go to the second cell and change the value of the angle variable from pi/2 to pi*2/3 and run that cell. You will see that the dependent cell gets automatically updated, including the plot.

Note that this was easier than the previous example, as we didn't need to define the angle as an Observable object; nor did we need to define listeners to allow re-rendering of the plot. Pluto makes it simple to modify the underlying data and get the visualization to update accordingly.

We need to consider that using Plots to render the plot and Pluto or Observables to trigger its re-rendering has limitations. In particular, as we are generating an entirely new plot on each update, we need to use a backend library that is fast enough to make the most of the interaction. Plotting lots of elements or using a slow backend can make our interactive visualization laggy. Thankfully, the default *GR backend* is one of the fastest. Also, if you have many points, inlining a vector image as **Scalable Vector Graphics (SVG)** on a notebook can make it unresponsive, but you can choose to inline a raster-graphic file format such as **Portable Network Graphics (PNG)** to speed up the rendering for smooth interaction in those cases. Luckily, Pluto does this automatically when we use Plots and the *GR backend*, as it decides to render a PNG figure rather than an SVG every time we have more than 8,000 data points. If your plot has a low number of data points, you can also take advantage of the extra interactivity offered by the Plotly backend. Plotly renders an interactive SVG image, so its speed depends on the number of visual elements it needs to show. Currently, Pluto does not support the PlotlyJS backend because PlotlyJS depends on the WebIO library.

Besides the limitations mentioned previously, Pluto and Plots are an excellent combination to create an interactive visualization that reacts to changes in the input data. As we can use arbitrary Julia code to modify the data, the options are endless. What's more, the PlutoUI library offers us a set of widgets, so notebooks users can change the data without interacting with the code but instead through visual elements.

Creating interfaces with PlutoUI

Pluto exports the @bind macro, which allows you to bind an *HTML* object, particularly an *input tag*, to a Julia variable. PlutoUI offers a more Julian way of writing those HTML objects, so you don't need to write HTML code. It provides lots of widgets, from sliders and buttons to color pickers. Let's create a Pluto notebook, as in the previous example, but use a slider to define the angle value. We will also try to hide the code to create a nice and clean interactive visualization. Proceed as follows:

1. Open a new Pluto notebook.

2. Execute using `Plots, PlutoUI` in the first cell.

3. Create a new cell below and execute the following code:

```
@bind angle PlutoUI.Slider(0:0.01:2pi, default=pi/2, show_
value=true)
```

That code has created a `Slider` object starting in the `pi/2` value, as indicated by the `default` keyword argument. The slider can choose values from the range used on the first positional argument. `Pluto` assigns the current value to the `angle` variable when you move the slider, which triggers any cell depending on that variable.

4. Create a new cell below and execute the following code:

```
begin
    angles = range(0, angle, length=100)
    plot(cos.(angles), sin.(angles),
        ratio=:equal,
        xlims=(-1.5, 1.5), ylims=(-1.5, 1.5),
        legend=:none,
        framestyle=:none)
end
```

Here, we have used `begin` and `end` to create points and plot them in the same cell. As `Pluto` shows the output of the last expression, only the plot will be inline, and we can later hide all the source code.

5. Move the slider; you will see that the plot updates accordingly.

6. Replace the code in the second cell, the one *defining the slider*, with the following code and execute it:

```
md"""
angle $(@bind angle PlutoUI.Slider(0:0.01:2pi, default=pi/2,
show_value=true))
"""
```

That creates a *Markdown string*, in which we have used `$(...)` to interpolate the code using the `@bind` macro to link the `Slider` object and the `angle` variable. Interpolating `PlutoUI` widgets into Markdown strings is an excellent way to add context to them.

Hide the source code of all the cells by clicking on the **Show/hide code** eye icon that appears at the left side of each cell when the mouse pointer is over it.

These steps create an interactive plot, as we can see in the following screenshot:

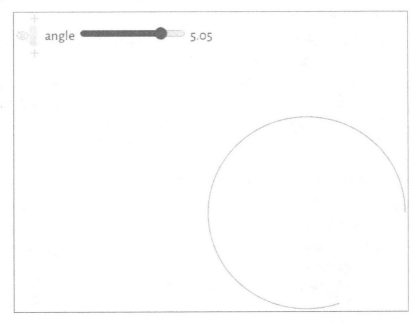

Figure 3.3 – Interactive plot using PlutoUI slider

Great! You now know how to use `Pluto` and `Plots` to create interactive plots that don't require the notebook user to modify the code. The following section will discuss how to use `Makie` and the `WGLMakie` backend to get similar behavior.

Interactive plotting with WGLMakie and Pluto

`Makie` can use `Observables` as input, updating only the changed figure's elements when the `Observable` value changes. Therefore, it doesn't need to re-render the entire plot, thus avoiding one of the limitations of `Plots`, but it also means that we need some extra work to make `Makie` work on the reactive framework of `Pluto`. In this section, we will take advantage of this `Makie` feature from `Pluto` while profiting from the interactivity offered by its `WGLMakie` backend. We can achieve this using `JSServe` or `PlutoUI` widgets, and we will explore the latter by reimplementing the previous example using `WGLMakie` instead of `Plots`. Proceed as follows:

1. Open a new `Pluto` notebook.

2. In the first cell of the notebook, execute the following code:

```
begin
    using JSServe
```

```
    Page()
end
```

The output of `Page` must be on the first cell of the notebook.

3. Create a new cell below, and execute `using WGLMakie, PlutoUI` to load both packages.

4. Create a new cell and execute the following code:

```
points = WGLMakie.Observable(Point2[]);
```

That creates `Observable` points wrapping `Point2[]`, an empty vector of `Point2` objects. The semicolon at the end hides the output of the cell.

5. In a new cell, execute the following code:

```
md"""
angle $(@bind angle PlutoUI.Slider(0:0.01:2pi, default=pi/2,
show_value=true))
"""
```

This, as before, creates a slider inside a Markdown string that updates the value of the `angle` variable.

6. Create a new cell and execute the following code:

```
points[] = [Point2(cos(a), sin(a)) for a in range(0, angle,
length=100)];
```

This updates the value of the `points` observable with the current points that draw the arc from 0 to our selected angle. Each time we move the slider, changing the value of angles, `Pluto` runs this cell, updating the `Observable` value.

7. In a new cell, execute the following code to draw our arc:

```
lines(
    points,
    axis=(aspect=DataAspect(), limits=(-1.5, 1.5, -1.5, 1.5)),
    figure=(resolution=(500, 500),)
    )
```

Here, we used the `points` variable, but since the cell doesn't contain any other variable from our notebook, `Pluto` would not re-evaluate this cell unless we were changing the `points` variable. Thankfully, `Pluto` doesn't trigger re-execution when the wrapped value of an observable is updated using `[]`. This behavior is helpful here, as we do not want to re-render the entire plot, only the elements that change.

8. Move the slider to see the change reflected in the plot. In this notebook, when we move the slider, the `angle` variable changes its value. `Pluto` then automatically runs the cell using the `angle` value, changing the value in the `points` Observable. Finally, the change in the points-wrapped value triggers `Makie` to update the needed plot elements.

9. Hide the source code of the cells to get a cleaner interface by clicking on the **Show/hide code** eye icon at the left side of the cell.

Now that we can create interactive visualizations using `Pluto` notebooks, let's move on to `IJulia`.

Interactive visualization with Jupyter

As with `Pluto`, we can rely on plotting packages wrapping *JavaScript* libraries to obtain interactive plots, but we can also use the power of `Observables` to interact even with static plotting ones—in particular, by taking advantage of the widgets offered by the `Interact` package. `Interact` is built around `Observables` to define the interactive behavior and `WebIO` to link Julia and the web. `Interact` can be used on both `Jupyter` and standalone web pages. Here, we will learn how to use it with `Jupyter`.

Setting up Jupyter for Interact

First, you will need to install the `WebIO`, `Interact`, and `IJulia` packages and set up your `Jupyter` environment to use `WebIO`. Proceed as follows:

1. Go to the terminal, type `julia`, and press *Enter* to open the Julia REPL.

2. Press the] key to enter **Pkg** mode, type `add WebIO Interact IJulia`, and press *Enter* to install the needed packages.

3. Type `using IJulia` and press *Enter* to load the `IJulia` package.

4. Execute `IJulia.JUPYTER` and see whether `.julia/conda` is part of the path. If it is part of the `Jupyter` path, you are using `Conda`, and the next step will depend on that.

5. If you are using `Conda`, then execute the following command on the Julia REPL:

    ```
    IJulia.Conda.pip _ interop(true)
    IJulia.Conda.pip("install",  "webio _ jupyter _ extension")
    ```

If you are using a *local installation* of Jupyter rather than Conda, then run the following command on your system terminal:

```
python3 -m pip show jupyter
```

Note that you should use python3.exe rather than python3 on *Windows*.

If the path indicated in **Location** is your user home, then you should run the following command with the –user flag:

```
python3 -m pip install --upgrade --user webio_jupyter_
extension
```

Otherwise, you should run the following command, which needs sudo access on *Unix* systems, to perform the system-wide installation:

```
python3 -m pip install --upgrade webio_jupyter_extension
```

Great! We now have everything we need to use Interact in Jupyter. So, let's use it!

Taking advantage of Interact

Now we have everything set up, let's plot our favorite arc, using Interact to change the angle value on Jupyter. Interact exports the @manipulate macro, which we can use to create multiple widgets. Interact is based on Observables, and the result type of the @manipulate macro is, in fact, an Observable object. The syntax for this macro resembles that of a for loop. It can define multiple variables, and the rendered widgets will depend on the variable types. For example, if the default value is a range of numbers, you will have a slider as a widget, while a Boolean will create a checkbox. Let's explore Interact with the following example:

1. Open a Julia REPL and execute using IJulia.
2. Execute notebook(dir=".") to open Jupyter in the current folder.
3. Create a new Julia notebook by clicking on the **New** button.
4. Type using Plots, Interact in the first cell of the notebook, and press *Shift + Enter* to execute that cell while creating a new one below.
5. Execute the following code in the new cell:

```
@manipulate for angle=0:0.01:2pi
    angles = range(0, angle, length=100)
    x = cos.(angles)
    y = sin.(angles)
    plot(
        x, y,
```

```
                ratio=:equal,
                xlims=(-1.5,  1.5),  ylims=(-1.5,  1.5),
                legend=:none,
                framestyle=:none)
    end
```

We have created an interactive plot using `Plots` and the default *GR backend*, reacting to the angle value selected with the angle slider (see *Figure 3.4*). The `@manipulate` macro syntax is simple; we need to define variables after the `for` loop and then use those variables in the code in the *body expression*. Here, we have defined an `angle` variable with the range `0:0.01:2pi`. Therefore, `Interact` shows that variable as a slider. Then, we use the `angle` variable in the body expression to create an `angles` list defining the *x* and *y* coordinates of the arc's points.

Those steps generate an interactive visualization, as we see in the following screenshot:

Figure 3.4 – Interactive plot in Jupyter using Interact

At this point, you can create interactive visualizations with `Plots` and `IJulia`, thanks to the use of the `Interact` library. In this section, we have also learned how to make them using `Pluto` and `PlutoUI`. Now, let's move on to the next section, where we will discuss Julia's different tools to share interactive visualizations.

Sharing your interactive visualizations

Up to this moment, we have seen different tools that Julia offers us to create interactive visualizations. Now, we are going to discuss how we can share them. We will also showcase some Julia libraries to create shareable interactive dashboards. The way we can share an interactive visualization depends on the tools we have used to create them. But first, let's start with the most basic way to share Julia projects.

Creating a Julia application

Creating and distributing a Julia application is an excellent way to share your interactive Julia visualizations; this section will give tips on how to achieve that. A **Julia application** is simply a folder with a specific structure and project environment with `Project.toml` and `Manifest.toml` files. We have seen how to create a project environment in *Chapter 1, An Introduction to Julia for Data Visualization and Analysis*. Therefore, we will describe the folder structure here. Julia applications usually contain the following subfolders:

- `src`: This folder contains the Julia source code of your application.

- `test`: This should include a `runtests.jl` file, which should have the test code for your application. It can also contain a project environment, with a `Project.toml` file declaring test dependencies.

- `docs`: This folder contains the source code to create application documentation, generally using the `Documenter` package.

Including a `README.md` Markdown file with indications needed to install and launch your Julia application is also sensitive. It is common and recommended to use *Git* to manage your Julia application. So, if your application is a Git repository, it is easy for Julia to install it using the package manager and its web link.

The most straightforward way to create a folder structure for a new application is by taking advantage of the `PkgTemplates` package. It is possible to use `PkgTemplates` because Julia applications and packages share the same folder structure; however, avoid including the `Manifest.toml` file in the `.gitignore` file. While packages should avoid committing the `Manifest` file, applications must ensure that all users have identical dependency versions by providing their manifest.

Having an application that Julia can install through the package manager could be the first step. You can also distribute that application using containers that contain both Julia and your application if you want—for example, using *Docker* or *Singularity*.

Creating a Julia application could be the basis for distributing most of your interactive visualizations. GLMakie interactive applications are good candidates to be distributed as Julia applications. You can also create a Julia application to distribute web server-like applications that users can install and serve locally—for example, you can distribute Interact standalone web pages or Dash and Stipple dashboards as Julia applications.

Sharing Pluto notebooks

As a Pluto notebook file contains its project environment, you can distribute it using a Git repository containing the notebook and a README.md file with indications on how to open it and run it. Also, in *Chapter 1*, *An Introduction to Julia for Data Visualization and Analysis*, we mentioned the advantages of sharing *static HTML* output. The exported HTML document contains the notebook file and the notebook's project environment. While the static HTML output is not interactive, you can extract the notebook file to run it locally by clicking on the **Edit or run this notebook** button. Additionally, that button also offers an option to run the notebook on the cloud using Binder. Binder is a free and open source service to run Jupyter and Pluto notebooks on the cloud. However, while Binder allows you to run the application without a local Julia installation, it can be slow and take some minutes to load.

If you want to publish your Pluto notebook as an interactive website where visitors can interact with PlutoUI widgets without modifying the notebook source code, you need to use the PlutoSliderServer package.

Publishing JavaScript plots

We have seen that *JavaScript* plotting libraries can produce interactive plots; one example of these packages is the Plotly backend of Plots. Those interactive plots can be saved and distributed as interactive HTML documents. In those cases, using the Weave package for literate programming can help to give the context the plots need to tell their stories. Weave allows interactive visualizations created with the Plots package and the Plotly backend when you export the document as an HTML file.

You can use also include your interactive plots on static websites. For example, the Franklin package allows you to add interactive Plotly figures to your website. Also, the Documenter package can inline interactive Gadfly and VegaLite plots out of the box.

Creating applications to serve interactive plots

Julia offers excellent utilities to create web servers, from the simple Mux package to the full stack Genie framework. What's more, you can create complex interactive dashboards in Julia using Stipple or Dash.

In the Pluto example using WGLMakie, we have seen that the *JavaScript* backend needed JSServe to work. JSServe is a library that links Julia and JavaScript and a server that you can use to create interactive dashboards. JSServer offers an attractive feature; it can store a state map of all the slider's values so that you can export a static HTML page where the slider is interactive without needing a Julia process.

The Interact package allows you to create a standalone web page using the Mux and WebIO packages. Let's reproduce the Interact and Jupyter example from the *Taking advantage of Interact* section of this chapter as a web page, as follows:

1. Go to the terminal and execute julia -e 'mkdir("arc")' to create an arc folder for our application.

2. Type julia in the same location and press *Enter* to open the Julia REPL.

3. Press the] key to enter the Pkg mode and execute activate arc, creating a new project environment on the arc folder.

4. Execute add Plots Interact Mux to add the needed packages for our application.

5. Press *Ctrl + D* to exit Julia.

6. Create a Julia file named interact_example.jl in the arc folder with the following content:

```
using Plots, Interact, Mux

app = @manipulate for angle=0:0.01:2pi
    angles = range(0, angle, length=100)
    x = cos.(angles)
    y = sin.(angles)
    plot(
        x, y,
        ratio=:equal,
        xlims=(-1.5, 1.5), ylims=(-1.5, 1.5),
        legend=:none,
        framestyle=:none)
```

```
end

WebIO.webio_serve(page("/", req -> app), 8085)
```

The code is almost identical to the example we executed using `Jupyter`, except we are now loading `Mux` to create a simple web server. We also stored the output of the `@manipulate` macro in the `app` variable. We later used that variable to make the web page, which `Mux` then served using `webio_serve`.

7. Go to the `arc` folder in the terminal and execute the following command:

```
julia --project=. -i interact_example.jl
```

The first time you execute this command, it may take some minutes until you see the Julia REPL, depending on the number of packages Julia needs to precompile. We used the `-i` flag to start an interactive Julia session, avoiding the Julia process to finish after the script run. Otherwise, we will lose the connection between Julia and the web page.

8. Access `http://localhost:8085/` on your web browser to see your interactive visualization. The resulting web page resembles the output of the second cell in *Figure 3.4*.

Great! You can now distribute your interactive visualization created with `Interact` as `Jupyter` notebooks or standalone web pages.

As we previously mentioned, you can create complex dashboards in Julia using `Stipple` or `Dash`. `Stipple`, based on the `Genie` framework, is relatively new and specific to Julia. It has excellent support for `Plotly` plots, thanks to the `StipplePlotly` package. However, the most basic plotting library for `Stipple` is `StippleCharts`, based on the `ApexCharts` interactive JavaScript library. When using `Stipple`, you can take advantage of the widgets and web elements, including tables and banners, offered by the `StippleUI` package.

`Dash` is a well-established open source framework for building dashboards on Julia, Python, and R. `Dash` uses `Plotly` for visualizations. Therefore, we can also create dashboards using `Plots` and the `Plotly` or `PlotlyJS` backends, but the interactivity between the plots and the widgets works differently from what we have seen: you need to register *callbacks* between `Dash` components. Let's create a dashboard for our arc example using `Plots` and the `Plotly` backend, as follows:

1. Open the Julia REPL.
2. Press `]` to enter the `Pkg` mode and execute `activate --temp` to create a temporal project environment.

3.   In Pkg mode, run the following code:

```
add Plots@1.22.3 PlotlyBase@0.8.18
add Dash@0.1.6 DashHtmlComponents@1.1.4
add DashCoreComponents@1.17.1
```

This will install a specific version of each needed package on the temporal environment. Note that this code example could not work with the latest Dash versions.

4.   Press *Backspace* to go out of the Pkg mode.

5.   Execute the following code in the Julia REPL to load Plots and the Dash libraries:

```
using Plots
using Dash, DashHtmlComponents, DashCoreComponents
```

6.   Execute plotly() to use the Plotly backend.

7.   Create a new Dash application by running app = dash().

8.   Run the following code:

```
app.layout = html _ div() do
    dcc _ slider(
        id="angle _ selector",
        min=0,
        max=2pi,
        step=0.01,
        value=pi/2,
    ),
    dcc _ graph(id="plot")
end
```

Here, we are creating an HTML page but using Julia objects. First, we have assigned an html_div object to the Dash application layout. Note that the html_div function, and other Dash components, can take a function as the first argument, allowing us to use the do syntax. The div element we have created has two elements: a slider and a plot. We created the former using the dcc_slider function and the latter thanks to the dcc_graph function. In those functions, dcc stands for **Dash Core Components**.

9. Run the following code:

```
callback!(
        app,
        Output("plot", "figure"),
        Input("angle_selector", "value")) do angle
    angles = range(0, angle, length=100)
    x = cos.(angles)
    y = sin.(angles)
    fig = plot(
        x, y,
        size=(500, 500),
        xlims=(-1.5, 1.5), ylims=(-1.5, 1.5),
        legend=:none,
        framestyle=:none)
    Plots.plotlybase_syncplot(fig)
    Plots.backend_object(fig)
end
```

Here, we have used the callback! function to create a link between the slider and the plot. In particular, the slider's value will be the Input for the plot figure, the Output. The code inside the do block creates a plot figure object. You will note that the code is identical to the one making the plot in the Interact example. However, here, we are determining the size of the plot rather than the ratio, as Plotly will ignore the latter. Also, rather than returning a Plots object, we need to return a Plotly one. To achieve that, we have used the plotlybase_ syncplot and backend_object functions. If we use the PlotlyJS backend, we need to call the plotlyjs_syncplot function instead of plotlybase_ syncplot.

10. Run run_server(app) to serve our application.

11. Copy the address that run_server shows after **Listening on:**, then paste it into your web browser's address bar to access the application.

The following screenshot shows an example of the Dash application we have just created using Dash and the interactive Plotly backend of Plots:

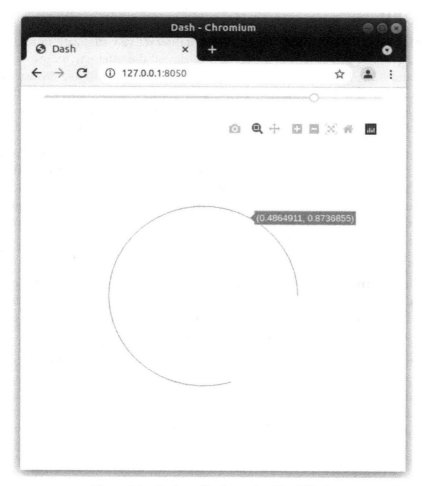

Figure 3.5 – Dash application using Plots' Plotly

Great! At this point, you have the minimum knowledge needed for creating and sharing dashboards and interactive visualizations in Julia. However, we have only scratched the surface in this chapter. If you want to develop complex applications, you can choose one of these tools and further learn it by reading its documentation or exploring its examples.

Summary

This chapter has taught us the different behaviors of interactive visualizations and which plotting packages offer them. We have seen how to exploit the interactivity provided by the `Plotly` and `PlotlyJS` backends of `Plots`. We learned how to use `Observables`, `Interact`, or `Pluto` to gain interactivity when rendering `Plots` figures with the default *GR backend*. Then, we also explored the enhanced and fully customizable interactive behaviors that `Makie` offers us, including mouse interaction.

Using different Julia packages, we created the same interactive plot—an arc with a varying internal angle. That gave us a good idea about the differences and similarities of those libraries to achieve interactive plots. Therefore, you will now be able to choose the best-suited combination of packages for your interactive data visualizations. This chapter only skimmed all the different libraries' options, functions, and widgets; we recommend looking at their documentation for more details. Nevertheless, you can use the code examples in this chapter as a starting point to create interactive figures.

Finally, you have learned how to create and distribute interactive visualizations. We first discussed how to create interactive plots using `Pluto` and `IJulia` notebooks. We have learned how to develop `GLMakie` applications that you can later distribute as Julia applications, and we can now even create standalone web pages and dashboards.

While allowing complete control of different plot aspects through interactive visualization is sometimes needed, in other situations creating animations showing pre-selected states could be enough. In the next chapter, we are going to discuss how to create animations with `Plots` and `Makie`. We will see that `Makie` animations use the same `Observables` system we explored in this chapter for interactivity. The next chapter will also introduce a new package: `Javis`.

Further reading

You can learn more about the packages mentioned in this chapter by reading their documentation. You can find the documentation for dealing with `Makie` mouse and keyboard events at `https://makie.juliaplots.org/stable/documentation/events/`. Some of the libraries, such as `JSServe`, don't have extensive documentation, but they offer a series of examples on their Git repositories.

4
Creating Animations

Animations and interactive visualizations can tell stories that static plots cannot. Also, both are more engaging for the final audience. As shown in the previous chapter, interactive visualization can let us interact with a variable to visualize its impact, among other things. Instead, animations use the time dimension to encode that variable. The encoded variable can be time since animations are well suited to show how a process evolves through time. Another advantage of animations is that they are much easier to distribute than interactive visualizations.

In this chapter, we are going to learn how to create animations using `Plots` and `Makie`. We will also introduce `Javis`, a drawing library, which we will use to generate animations with Julia. Then, we will learn about the `Animations` package, which will help us move objects in our canvas. By the end of this chapter, you will know how to create and distribute simple animations using `Plots`, `Makie`, and `Javis`.

In this chapter, we're going to cover the following main topics:

- Easy animation using `Plots`
- Animating `Makie` plots
- Getting started with `Javis`

Technical requirements

For this chapter, you will need Julia, Pluto, and IJulia installed. You will also need a web browser with an internet connection. The Pluto and Jupyter notebooks for this chapter can be found in the Chapter04 folder of this book's GitHub repository: https://github.com/PacktPublishing/Interactive-Visualization-and-Plotting-with-Julia. You will also need to install the Plots package. This chapter also requires GLMakie, Animations, and Javis, but Pluto will automatically install them as required.

Easy animation using Plots

In this section, we will learn what animations are and how we can create them with Plots. We will also discuss the Animations package, which will help us interpolate between positions. We are also going to take advantage of Animations when working with Makie and Javis. But let's start with the most straightforward animation tool available in Julia: Plots.

Animations are simply a succession of slightly similar images in time that give the sensation of movement. Each of those images is a frame of the animation. The Plots package does precisely that; it provides an Animation object that you can add one frame at a time to. Each frame is a static PNG image that's created using Plots. It's up to us to generate each frame with the slight change that's needed to create a pleasant animation.

We can save animations in different formats. The Plots package offers the following format options:

- gif: **GIF** is the most popular and widespread file format for small animations. Almost all platforms, including old web browsers, support this format. You can even use it on social media platforms. GIF is also the only animation format that the Julia extension for **Visual Studio Code** (**VS Code**) can display out of the box. GIFs are usually low quality compared to the other options as they only support 256 colors and low **frames per second** (**FPS**). Because of its lossless compression, complex animations can lead to heavy files. Note that the first reproduction of a heavy GIF file could be slow.

- webm: Most modern web browsers support **WebM**. It offers higher quality and smaller file sizes than the previously mentioned *GIF* format.

- mp4: **MP4** is one of the most popular formats; therefore, almost all platforms support it. It is also helpful for sharing videos on social media. It offers small file sizes while keeping good quality. However, *WebM* files tend to be a little smaller than MP4 files while maintaining similar quality.

- mov: The **MOV** format, also known as the **QuickTime File Format (QTFF)**, is the least supported format in this list. It offers the highest quality, but it also generates the heaviest files.

For each of these formats, there is a function with the same name in lowercase. Those functions can take an Animation object and save it in that format. There is also a @gif macro, which can help you create a GIF animation quickly. When displayed using HTML, for example, inside Pluto or Jupyter notebooks, the MOV, MP4, and WebM formats offer a series of buttons to control video reproduction. However, support would depend on your operating system, web browser version, and available plugins. Note that you can also reproduce those animation files using movie players. In particular, the *VLC media player* supports all those formats while being free, cross-platform, and open source.

Let's explore this by creating a simple animation using Plots. In this example, we are going to move a point around the origin while following the unit circle. First, we will show you the most verbose approach to help you gain an idea of what Plots is doing. Later, we will learn how to do this more easily:

1. Open the Julia REPL and execute using IJulia; notebook() to open Jupyter.

2. Create a new Jupyter notebook.

3. Execute using Plots in the first cell to load the Plots package and the default GR backend; we do not need any other packages to create a Plots animation.

4. Execute the following code in a new cell:

```
function plot_frame(angle)
    scatter(
        [cos(angle)],
        [sin(angle)],
        ratio=:equal,
        xlims=(-1.5, 1.5), ylims=(-1.5, 1.5),
        legend=:none)
end
```

Here, we have created the function that we will use to create each frame of the animation. For the animation, we will slightly change the value of the internal angle for each frame and display a point at the coordinates where the arc ends. In other words, we will create a scatter plot with a single point located at the x and y coordinates, as determined by the cosine and the sine of a given angle. So, we use a vector with a single element for the x and y coordinates. As shown in the examples in *Chapter 3*, *Getting Interactive Plots with Julia*, we have set the limits for the axes so that they fix our canvas; otherwise, we would see a fixed point and changing axes. We have also set the ratio to :equal to ensure that we have a well-proportionated circle, and we have set the legend to :none so that we have a cleaner canvas.

5. Run the following code in an empty cell:

```
anim = Animation()
```

This creates an Animation object that, at the moment, has no frames. The Animation object contains the path to a temporal directory, where the frames will be stored. Great! Now, it is time to create the frames of our animation.

6. Create a new cell and execute the following code:

```
for angle in 0:0.05:2pi:
    plt = plot_frame(angle)
    frame(anim, plt)
end
```

The preceding code iterates from 0 to 2 times pi to get the value of the angle variable using small steps of 0.05. Therefore, we are creating the slight change we need between frames. For each iteration, we create the plot object containing the image for a frame. Then, we store the current frame in the animation using the frame function.

7. Execute anim in a new cell to see the Animation object. The frame field should now contain a vector of PNG images.

8. In an empty cell, execute gif(anim). This function will tell you the address of the temporal file containing the animation, and it will show the animated GIF inline in the notebook. The following screenshot shows what you will see after executing these last two steps; note that one of the frames of the animation has been frozen here:

```
In [5]:    1  anim
```

```
Out[5]:  Animation("/tmp/jl_PjvvJI", ["000001.png", "000002.png",
         "000003.png", "000004.png", "000005.png", "000006.png",
         "000007.png", "000008.png", "000009.png", "000010.png"  ...
         "000054.png", "000055.png", "000056.png", "000057.png",
         "000058.png", "000059.png", "000060.png", "000061.png",
         "000062.png", "000063.png"])
```

```
In [6]:    1  gif(anim)
```

Out[6]:

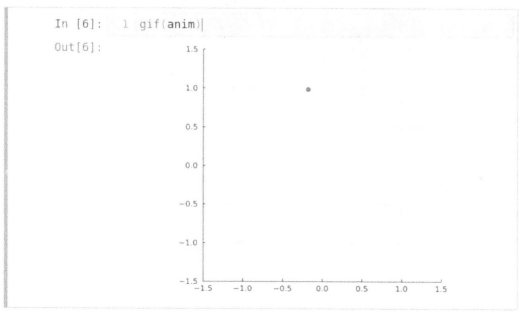

Figure 4.1 – Animated GIF inline on our Jupyter notebook

Perfect! You have created your first animation with the Plots package. Note that the gif function, similar to the mov, webm, and mp4 functions, can take a second position argument to specify the filename of the final animation. You must choose the *extension* based on the format; therefore, it should be equal to the function's name. These functions can also take the following keyword arguments:

- fps: This keyword argument determines the frame rate as the number of FPS; it defaults to 20.

- loop: This keyword argument takes an integer number indicating the number of times the player should reproduce the animation. The GIF format supports it well. The default value is 0, meaning to loop it infinitely. You can set this value to -1 to avoid looping the animation. If you set this to *n*, with *n* greater than 0, the player will play the animation *n + 1* times.

For example, `gif(anim, "first_animation.gif", fps=50, loop=-1)` will save the GIF as `first_animation.gif`, so the animation will be faster than the default one, and it will not loop.

We have created an animation in the most verbose way we can by explicitly populating an `Animation` object. We did this in Jupyter as doing the same in Pluto won't play nicely with reactivity. That's because the `frame` function modifies the `Animation` object in a way Pluto cannot figure out. So, the same example in Pluto should create the `Animation` object and call the `frame` function in the same cell by using a single code block that's been defined, thanks to the `begin` and `end` keywords. Thankfully, `Plots` offers a more convenient way to create animations that work nicely on all the editors – the `@animate` macro. Let's make a more straightforward version of our animation by using it:

1. Open a new Pluto notebook.

2. Execute `using Plots` in the first cell.

3. Create a new cell and execute the following code block:

```
function plot_frame(angle):
    scatter(
    [cos(angle)],
    [sin(angle)],
    ratio=:equal,
    xlims=(-1.5, 1.5), ylims=(-1.5, 1.5),
    legend=:none)
end
```

As we did previously, we have created a function that creates the image we want for each frame. We have done this to keep our code clean, but creating a dedicated function is not mandatory since you can create any plot inside the `for` loop.

4. Execute the following code in a new cell:

```
anim = @animate for angle in 0:0.05:2pi
    plot_frame(angle)
end
```

As you can see, we have created and populated the `Animation` object using the `@animate` macro in front of the `for` loop that renders each frame. We do not need to call the `frame` function now since the `@animate` macro automatically does this.

5. Create a new cell and execute `gif(anim)` to view our animation.

As you can see, everything has been more straightforward in this example. The `@animate` macro is the preferred way to create animations with `Plots`. It can also take some extra flags after the `for` loop block that could be helpful when you're creating animations:

- every: The `every` flag takes an integer, *n*, to save a frame every *n* iterations.

- when: The `when` flag takes a Boolean expression that will save a frame each time the expression is `true`.

For example, the following code uses the `@animate` macro and its `every` flag to create a faster but less smooth animation by taking fewer frames:

```
anim_fast = @animate for angle in 0:0.05:2pi
    plot_frame(angle)
end every 10
gif(anim_fast)
```

The following code uses the `when` flag of `@animate` to save frames, but only when angle is between 0 and `pi`:

```
anim_partial = @animate for angle in 0:0.05:2pi
    plot_frame(angle)
end when 0 <= angle <= pi
gif(anim_partial)
```

In these examples, we created an `Animation` object and then called the `gif` function to render it. This gave us a lot of freedom to choose how to render it as we can select the file format, the frame rate, and the looping behavior. However, this can be cumbersome for throwaway animations we want to see quickly. For those cases, `Plots` exports the `@gif` macro, which works like the `@animate` macro, but it returns a GIF file rather than an animation object.

Now, you know how to create animations with `Plots` and save them in different formats for distribution. We have created a simple animation, but you can animate more complex figures by taking advantage of all the available `Plots` features. In the next section, we will explore the `Animations` package, combined with `Plots`, to move objects around our canvas.

Using the Animations package

The Animations package helps us create animations by interpolating between different positions in time. You can control the interpolation of the relevant values using **easing functions**. These easing functions will determine the speed and acceleration of the movement in the animation. For example, the noease easing function will avoid any interpolation; all the animation frames will show the object at the initial position, except for the last one. The stepease function is similar, but it will let us choose when the position changes. The linear easing function will gradually change the object positions between frames without acceleration. For example, we can make a point rotate around the origin at the same speed, as in our previous animation, using the linear easing function to interpolate the angle values. Any other easing function will create an accelerated movement. Let's look at an example by using the Animations package and the polyio easing function to make our point go faster for angle values around pi:

1. Open a new Pluto notebook.

2. Execute using Plots, Animations in the first cell to load the libraries.

3. Create a new cell and execute the following code:

```
angle_animation = Animations.Animation(
    0, 0.0,
    polyio(3),
    1, 2pi
)
```

This creates an Animation object from the Animations package. We needed the Animations.Animation qualified name as Plots and Animations both export an Animation type. We created an animation that interpolates the angle value from 0.0 at time 0 to 2pi at time 1. We set that in the Animation constructor by passing the number pairs 0, 0.0 and 1, 2pi. Each pair creates a Keyframe object, where the first number determines the time and the second is the value to interpolate. This last value should be a floating-point number. If we explore the output of that cell, we will see that the frame field of the Animation object has a vector of Keyframe objects. We indicated the easing functions between the pair of numbers that are determining the keyframes. In this case, we used a polyio, which creates a polynomial easing. This function takes one parameter to indicate the power of the polynomial function, but not all the easing functions take parameters.

4. Execute times = 0:0.005:1 in a new cell; this creates our time variable. The range should include the time values of our keyframes.

5. In a new cell, execute the following code:

```
plot(times, angle_animation.(times),
    xlab="t", ylab="angle", legend=false)
```

In this way, we can visualize how the value of the angle changes as a function of our time variable. Here, we have used dot broadcasting to calculate the angle value for each time point. This is possible because the `Animation` object, when called as a function using a `Real` number, will return the value at that time point. The following screenshot shows the plot with the shape that was created by the `polyio` easing function. The function's slope indicates that the fastest speed will be at time 0.5:

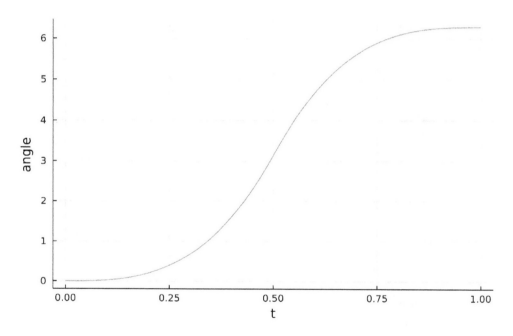

Figure 4.2 – A change in the angle value through time

6. Create a new cell and execute the following code:

```
anim = @animate for time in times
    angle = at(angle_animation, time)
    scatter([cos(angle)], [sin(angle)],
        xlims=(-1.5, 1.5), ylims=(-1.5, 1.5),
        ratio=:equal, legend=:none)
end
```

This code creates a `Plots Animation` object using the `@animate` macro. Inside the `for` loop, there's a call to `scatter`, which is the same as in the previous examples. The main difference is that we defined the angle value using the `at` function, which takes the interpolated value at a given time. In fact, `at(angle_animation, time)` is equivalent to `angle_animation(time)`.

7. Execute `gif(anim)` in a new cell to see the animation inline in the notebook.

We now have a rotating point with varying speeds! We determined the speed change using the `polyio` function. `Animations` also exports other easing functions, such as `polyin` and `polyout`. Some easing functions, such as `polyio`, can take keyword arguments. For example, you can use the `yoyo` easing function and its `n` argument to create a back-and-forth movement and determine the number of repetitions.

Now that we've learned how to create animations using `Plots` and learned how to use the `Animations` package to help us create complex movements, let's learn how to create animations using `Makie`.

Animating Makie plots

There are many ways to animate `Makie` plots; here, we will explore one of them. The general idea is similar to the one presented for `Plots`; we need to modify a plot to create the animation frames iteratively. To achieve that, `Makie` offers the `record` function, which can use `Figure`, `FigureAxisPlot`, or `Scene` to animate an iterator that returns the values changing through time and a function that uses these values to modify the plot. This function is the first argument to `record`, so we can use the `do` syntax to create it. Inside that function, you can rely on the `Observables` mechanism, which we discussed in *Chapter 3, Getting Interactive Plots with Julia*. The advantage of this is that `Makie` updates only parts of the plot that have changed in each frame, rather than rendering an entirely new figure, which is what `Plots` does. Let's explore this by creating a GIF animation of the rotating point using `GLMakie` and `Pluto`:

1. Create a new Pluto notebook.
2. Execute `using GLMakie` in the first cell to load `Makie` and its OpenGL-powered backend.
3. Create a new cell and execute `angle = Observable(0.0)`
4. In a new cell, run the following code:

    ```
    point = @lift Point2(cos($angle), sin($angle))
    ```

Here, we have used the `@lift` macro to generate a `point` observable that Julia will update each time the value of the `angle` observable changes.

5. Run the following code in a new cell:

```
plt = scatter(point,
      axis=(aspect=DataAspect(),
          limits=(-1.5, 1.5, -1.5, 1.5)))
```

The preceding code creates a `scatter` plot showing the `point` observable. We set the correct `aspect` to see the circular trajectory and the proper `limits` to ensure that the axis doesn't change through the animation. You will see the scatter plot inline in the notebook.

6. Add a new cell after that one and execute the following code:

```
anim = record(plt, "rotating.gif", 0:0.05:2pi) do value
    angle[] = value
end
```

The preceding code will create the animation by creating a frame for each value on the `0:0.05:2pi` iterator. The `record` function will pass each iteration value to the anonymous function we created with the `do` syntax. In this case, the anonymous function is only updating the value of the observable angle. This update causes `Makie` to change the necessary elements of the existing plot. You will see that `Makie` shows the changes in a GLFW window. The `record` function returns the path to the animation file. We have determined the name of that file in the call to the `record` function. The filename extension will select the output format; in this case, we are creating a GIF file. Let's use `LocalResource` from PlutoUI to inline the animation file in our notebook.

7. Create a new cell and execute `using PlutoUI`.

8. Execute `LocalResource(anim)` in a new cell to see the animation.

With that, we have created our first animation using `Makie` by taking advantage of the same system we learned how to use to create interactive plots. Here, we saved our animation in GIF format, as indicated by the `gif` extension. You can also store animation files using the previously mentioned extensions and formats; that is, `webm` and `mp4`. `Makie` cannot save animations in MOV format, but contrary to `Plots`, it also offers the **Matroska Multimedia Container** (**MKV**) format. This format is generated by default when you use `Makie`.

Now that we know how to create animations using `Makie`, let's learn how to use the `Javis` package for the same purpose.

Getting started with Javis

Javis is different from the previously mentioned options as it is not a plotting library like Plots or Makie. Javis only focuses on creating animations. It uses and re-exports objects from Luxor, a 2D drawing package based in Cairo. Here, we will quickly introduce the package by animating our rotation point. However, be aware that you can do much more with it. Let's learn how to create our animation with Pluto:

1. Create a new Pluto notebook and execute using Javis in the first cell.

2. Add a new cell and run the following code:

    ```
    function ground(args...)
        background("white")
        sethue("blue")
    end
    ```

    The preceding code creates a function that will take the Video object, the object to animate, and the current frame. Javis will give those inputs to all user-defined functions, so it is common to use args... for their arguments. This function calls background to set the background color and sethue to select the default color for our objects.

3. In a new cell, execute the following code:

    ```
    function create_point(args...)
        circle(O, 10, :fill)
    end
    ```

    This function will create the object to animate. In this case, we will create a circle with a radius of 10 at the origin. Javis defines O as an alias for Point(0.0, 0.0).

4. Run the following code in a new cell:

    ```
    begin
        video = Video(500, 500)
        Background(1:50, ground)
        point = Object(create_point, Point(100, 0))
        act!(point,
            Action(anim_rotate_around(2pi, 0.0, O)))
    end
    ```

Here, we have created a `Video` object for our animation. `Javis`'s functions and objects will modify the previously created video. Therefore, we need to execute them in the same Pluto cell, as Pluto cannot keep track of those changes. For example, here, we have set `Background` for our `Video` using the previously defined function. We have also determined that our video will have 50 frames using the `1:50` range. Then, we created a `point` object using the previously defined function and placed it at `Point(100, 0)`. Note that `Point(0.0, 0.0)` is the center of the canvas. Finally, we used the `act!` function to assign an `Action` to our object. In this case, we make our point rotate around the origin from angle `2pi` to `0.0`.

5. Create a new cell and execute `render(video; pathname="javis_animation.gif")` to save the animation. The file extension determines the file format; the available options are `gif` and `mp4`. You will see the animation inline in the Pluto notebook.

Great! We have created our first animation with `Javis`.

Summary

In this chapter, we learned how to create animations using `Plots`, `Makie`, and `Javis`. We have also learned how to use the `Animations` package to make complex movements with a few lines of code. While we have learned how to use the `Animations` package with `Plots`, you can also take advantage of it with `Makie` and `Javis`. Now, we can choose the correct format to distribute our animations in. You will find this new knowledge incredibly helpful while creating and sharing compelling animations. This, in turn, will help you visualize data that changes through time, make didactic animated figures, and draw attention to your plots.

In the next chapter, we will learn about the different adaptations of the grammar of graphics available in Julia.

Further reading

If you want to learn more about creating animations with Julia, please read the documentation for the packages we used in this chapter:

- Documentation for `Plots`: `http://docs.juliaplots.org/latest/animations/`.

- Documentation for `Makie`: `https://makie.juliaplots.org/stable/documentation/animation`.

- Documentation for the `Animations` package: `https://jkrumbiegel.com/Animations.jl/stable/`.

- In this chapter, we have only scratched the surface of what `Javis` can offer; you can explore it more by following the tutorials here: `https://wikunia.github.io/Javis.jl/stable/`.

Section 2 – Advanced Plot Types

In this section, we will explore predefined plot types scattered through different libraries in the Julia ecosystem. We will introduce packages and plots needed for data analysis and visualization. In particular, we will learn about DataFrames and the Julia packages inspired by *The Grammar of Graphics*, namely, `Gadfly`, `VegaLite`, and `AlgebraOfGraphics`. Next, we will introduce `StatsPlots` and explore how to create classical statistical plots with Julia. Then, we will learn how to work with graphs and networks in Julia and explore different plotting libraries for their visualization. After that, we will change the subject again to learn how to draw maps and visualize geographically distributed data. Finally, we will discover some Julia packages for plotting biological objects, such as sequences and protein structures.

This section comprises the following chapters:

- *Chapter 5, Introducing the Grammar of Graphics*
- *Chapter 6, Creating Statistical Plots*
- *Chapter 7, Visualizing Graphs*
- *Chapter 8, Visualizing Geographically Distributed Data*
- *Chapter 9, Plotting Biological Data*

5

Introducing the Grammar of Graphics

This chapter will introduce the Grammar of Graphics, which enables us to describe complex graphics simply and quickly. It is advantageous for creating statistical plots and performing data analysis tasks, as it allows for fast data exploration. Learning the basics of the Grammar of Graphics will allow you to catch up quickly with the numerous packages that implement some version of it. In particular, this chapter will teach you about the most popular Julia packages that use some variation of the Grammar of Graphics: `Gadfly`, `VegaLite`, and `AlgebraOfGraphics`.

This chapter will also introduce a fundamental tool for data analysis – the `DataFrames` package. We will see that the Julia libraries that use the Grammar of Graphics are tightly tied to this tabular representation of data. We will also discuss the correct data layout by introducing the concept of tidy data tables. This will significantly help you perform basic data analysis tasks in Julia and transform data for plotting purposes.

In this chapter, we're going to cover the following main topics:

- Introducing the Grammar of Graphics
- Working with DataFrames and tidy data

- Exploring data with Gadfly
- Introducing the Grammar of Interactive Graphics with VegaLite
- Makie's Algebra of Graphics

Technical requirements

You will need Julia, Pluto, a web browser, and an internet connection for this chapter. The Pluto notebooks that contain the code examples and their HTML versions that show the plot outputs can be found in the `Chapter05` folder of this book's GitHub repository: `https://github.com/PacktPublishing/Interactive-Visualization-and-Plotting-with-Julia`.

Introducing the Grammar of Graphics

Wilkinson's book, *The Grammar of Graphics*, defined grammar to express the underlying structure of statistical graphs. This allowed us to go from a fixed set of named graphs, such as scatter or line plots, to a whole range of more complex visualizations. This was possible since grammar sets up the rules to combine different components to create new graphs. The R language adaptation of this grammar, described in Wickham's article, *A Layered Grammar of Graphics*, has been the most influential to the Julia ecosystem. This section will discuss the main aspects of this grammar so that we can understand their adaptations to the Julia language.

The main components of a graph, according to the *Layered Grammar of Graphics*, are as follows:

- **Data:** This is a critical part of a plot that's able to turn an abstract graph specification into a concrete instance for that specific data. We will discuss data and its representation in more depth in the next section.

- **Mapping:** Mappings allow us to link data variables to **aesthetics**, which are visual features of the graph, such as position, color, and shape. These aesthetics that determine the appearance of the geometric objects are also known as visual **channels**.

- **Geometry:** There are geometric objects, also known as **marks**, such as points, lines, and bars that determine the plot type. For example, we can create a scatter plot by using points as our geometry. Different geometries allow for different aesthetics or channels. For example, a point geometry can have x and y positions, shape, size, color, and alpha transparency as aesthetics.

- **Statistics**: A statistical transformation is used on the input dataset to return new data variables.

- **Scale**: Scales control and keep track of the mapping between an input variable and its associated aesthetics. Therefore, they determine how the axes and legends evidence those mappings.

- **Facet**: This grammar element creates **small multiples**, also known as **trellis plots**, that use the same scales and axes while showing different subsets of the input data (see *Figure 5.5* for an example).

There are more grammar components, but these are the main ones to understand when it comes to the Julia ecosystem. An interesting aspect of these components is that they are independent. Consequently, we can combine them modularly to create new plots and quickly explore alternative visualizations.

In A *Layered Grammar of Graphics*, it's stated that we can combine multiple components into a **layer**. In particular, each layer is comprised of some data, a geometry, a mapping between the data and the geometry aesthetics, and a statistical transformation. That is enough to produce a plot. What's more, we can pile different layers that usually share the same data in a single visualization. For example, you can create a scatter plot with a regression line by combining a layer that uses the point geometry with another that uses the line geometry and the appropriate statistical transformation. We will see examples of this later in this chapter.

One of the advantages of A *Layered Grammar of Graphics* is its use of sensible **defaults**. This allows you to describe plots without requiring the full specification. Instead, you can create a concise graphs specification by omitting some grammar parts to rely on the plotting library defaults. While that allows for quick data exploration, it can make customizing plots more cumbersome.

The Grammar of Graphics includes data **transformations** as one of the grammar components. However, A *Layered Grammar of Graphics* does not incorporate that into the grammar as R can easily manipulate data. The same happens with Julia; there is no need to include data manipulation in the plotting grammar as Julia is a perfect tool for data wrangling. There is, however, an exception; the **Grammar of Interactive Graphics** of **Vega-Lite** includes data transformations so that it can apply them as the user interacts with the plot. We will explore that extension of the grammar in the *Introducing the Grammar of Interactive Graphics with VegaLite* section.

With that, we've looked at the main elements of *The Grammar of Graphics*. However, before we see this grammar in action, let's learn how to work with data in Julia.

Working with DataFrames and tidy data

To work within the grammar of graphics, we need data. But we do not need any data; we need **tidy data**. Tidy data is data that's been arranged in a tabular way, in which each row of the table represents an observation, and each column represents a variable. This layout is essential as we usually want to map data variables, and therefore columns, to geometry aesthetics. Usually, we use the DataFrame data structure to represent and store this kind of data. The `DataFrames` package defines this structure for the Julia language and exports many valuable functions to work with it. We are going to explore this package in this section.

DataFrames are usually stored using text files in **Comma-Separated Values** (**CSV**) format; therefore, we typically need the CSV package to load them. There is also a series of helpful datasets stored in the `RDatasets` and `VegaDatasets` packages that we will use for demonstration purposes throughout this book. Now, let's learn how to load an example dataset from `RDatasets` and save it using the CSV package:

1. Create a folder to store the code examples for this section.
2. Create a new Pluto notebook and save it in the previously created folder. Call it `save_iris_dataframe.jl`.
3. Execute `using RDatasets` to load the package.
4. Create a new cell and run the following code:

   ```
   iris = dataset("datasets", "iris")
   ```

   Here, we have used the `dataset` function to load the famous Iris dataset from the `RDatasets` package. The `dataset` function from the `RDatasets` package takes an R package name and a dataset name as arguments. You will see that Pluto shows a table of 150 rows and five columns after executing the preceding code (you can see a similar table in *Figure 5.1*). This dataset is a classic example of tidy data where each row represents an observation – in this case, an Iris flower. Each column represents a variable that's been measured for those flowers. The first four columns are quantitative variables that store the length and width of petals and sepals. The fifth column is a categorical variable that indicates the species name of the measured Iris flower.

5. In a new cell, run the following code:

   ```
   typeof(iris)
   ```

You will see that `iris` is a DataFrame from the `DataFrames` package. The `RDatasets` package uses the `Reexport` package to re-export all the exported names from the `DataFrames` package. So, you can use all the types and methods from the `DataFrames` package with it. Let's save this dataset into a CSV file.

6. Execute `using CSV` in a new cell.

7. Create a new cell and run the following command:

    ```
    CSV.write("iris_dataset.csv", iris)
    ```

    The preceding command uses the `write` function from the `CSV` package to save the `iris` DataFrame in the `iris_dataset.csv` file using the CSV format.

8. Close the notebook.

With that, we have learned how to load a DataFrame from the `RDatasets` package and save a DataFrame to disk in CSV format. Now, let's learn how to load a DataFrame from a CSV file and look at some helpful operations we can perform on that data structure on a new notebook:

1. Create a new Pluto notebook in the same folder where the `save_iris_dataframe.jl` notebook is and name the new one `describe_iris_dataframe.jl`.

2. Execute `using CSV, DataFrames` in the first cell.

3. Run the following code in a new cell:

    ```
    iris = CSV.File("iris_dataset.csv") |> DataFrame
    ```

    This code reads the table in the `iris_dataset.csv` file we saved with the `save_iris_dataframe.jl` notebook and stores it in the `iris` variable as a DataFrame from the `DataFrames` package, as shown in the following screenshot:

iris =

		SepalLength	SepalWidth	PetalLength	PetalWidth	Species
		Float64	Float64	Float64	Float64	String15
	1	5.1	3.5	1.4	0.2	"setosa"
	2	4.9	3.0	1.4	0.2	"setosa"
	3	4.7	3.2	1.3	0.2	"setosa"
	4	4.6	3.1	1.5	0.2	"setosa"
	5	5.0	3.6	1.4	0.2	"setosa"
	6	5.4	3.9	1.7	0.4	"setosa"
	7	4.6	3.4	1.4	0.3	"setosa"
	8	5.0	3.4	1.5	0.2	"setosa"
	9	4.4	2.9	1.4	0.2	"setosa"
	10	4.9	3.1	1.5	0.1	"setosa"
	⋮ more					
	150	5.9	3.0	5.1	1.8	"virginica"

```
iris = CSV.File("iris_dataset.csv") |> DataFrame
```

Figure 5.1 – DataFrame containing the Iris dataset in Pluto

Here, we can see that Pluto has a particular way of showing DataFrames. To begin, if the mouse is over the table, Pluto will show the data type of the different columns is displayed in gray. In this case, the numerical columns use Float64, and the species names are stored using an InlineString value of 15 code units from the WeakRefStrings package. You will also see that Pluto only shows the first 10 rows of the table and the last one. However, you can display more rows by clicking on the ⋮ **more** button before the last row.

There are some exciting things in the code that we have run in this step. First, File is a data type from CSV that allows us to read a file in CSV format. Therefore, to work with a DataFrame, we need to convert this File object using the DataFrame constructor. Here, we used the |> operator, a pipe, to input the File object into the DataFrame constructor. Therefore, the code is equivalent to the following:

```
iris = DataFrame(CSV.File("iris_dataset.csv"))
```

You can use either syntax, but the pipe operator is pretty popular for chaining operations when performing data wrangling activities. So, it is nice to get used to it.

4. In a new cell, execute the following command:

    ```
    describe(iris)
    ```

    The `describe` function will return a DataFrame that summarizes the variables in our dataset – that is, the columns in the input DataFrame. The function will show the name, stored data type, summary statistics, and the number of missing values for each column.

5. Execute `nrow(iris)` in a new cell to see the number of observations in our dataset. In this case, the Iris dataset has measures for 150 Iris flowers. If you want to know the number of columns, you can execute the `ncol` function instead.

6. In a new cell, execute the following line of code:

    ```
    unique(iris.Species)
    ```

    There are many ways to access the columns of a DataFrame; here, we have used the dot syntax to access the `Species` column of the `iris` DataFrame. Then, we showed the unique values on that column. We can see that this dataset contains three species of Iris flowers.

7. Run the following code in a new cell:

    ```
    species = groupby(iris, :Species)
    ```

    The `groupby` function takes a DataFrame and a column name or a vector of column names. We used a `Symbol` type to indicate the column name, but you can also use a `String` or an `Int` type to indicate the column's position. The `groupby` function returns a `GroupedDataFrame` that you can iterate to get a subset of the input DataFrame for each unique combination of values on the indicated columns. In this case, each group contains the observations for a given species.

8. In a new cell, run the following code:

    ```
    combine(species, nrow)
    ```

    Here, we can see that we have 50 observations or rows for each species. The `combine` function can take a `GroupedDataFrame`, apply a series of operations, and create a new DataFrame with the results. In this case, we calculated the number of rows of each group and returned a new DataFrame with those values. Of course, you can apply more complicated operations.

9. Create a new cell and execute the following code:

    ```
    combine(species, :SepalLength => minimum =>
    :MinSepalLength)
    ```

The preceding code uses the `DataFrames` specification to apply a function to a dataset. It consists of a column selector, the `:SepalLength` symbol, a function, `minimum`, and the name for the output column, `:MinSepalLength`. These are all linked using `=>`, the Julia pair operator.

10. Create another cell and execute the following code:

```
colnames = names(iris)
```

The `names` function will return a vector containing the column names in the DataFrame.

11. Execute `numeric_columns = colnames[1:4]` in a new cell to get a vector containing only our dataset's first four column names.

12. Execute the following code line in a new cell:

```
transform(iris, numeric_columns .=> (x -> 10x) .=>
numeric_columns)
```

This transformation allows us to convert our measures from centimeters into millimeters. We can do this using the `select` or `transform` functions. Here, we have chosen the `transform` function, which returns a DataFrame with the input columns plus the new ones, as `select` will drop the unselected `Species` column. In the *specification*, we used the same vector of column names to indicate the input columns and the names of the new ones; therefore, we are replacing them. Note that since we are applying our anonymous function to multiple columns, we needed to use `.=>`, with the dot syntax for broadcasting, instead of `=>`. The parentheses around the anonymous function are necessary; otherwise, Julia will consider the second `.=>` operator as part of the function.

13. In a new cell, execute the following code block:

```
begin
    iris[!, :Sample] = 1:nrow(iris)
    long = stack(iris, Not([:Species, :Sample]))
end
```

Here, we created a column to identify the samples, and then we went from our wide DataFrame to a long one using the `stack` function. We executed both commands in a single code block as Pluto cannot track mutating operations. In this example, we implicitly used the `setindex!` mutating function to create the `:Sample` column.

To create a new column, we used the indexing syntax, where we used a bang, !, instead of a colon, :, to index the rows. In DataFrames, using a colon means making a copy, so we need to use a bang to modify the DataFrame. Then, we selected the nonexistent :Sample column to create it by assigning it a range with as many numbers as there are rows in the DataFrame.

Once we created our identifier for each sample, we used the stack function to reshape our tidy dataset into a long non-tidy version of it. The stack function takes the columns to stack as the second positional argument. In this case, we gave all the columns that are not Species or Sample. We took advantage of the DataFrames Not selector for that. By default, the stack function considers all the remaining columns as identifiers. The DataFrame keeps the identifiers column during stacking while repeating their values accordingly to create a long-format DataFrame.

14. Create a new cell and execute the following code block in it:

```
begin
    long.Dimension = [
        replace(x, r"Sepal|Petal" => "")
        for x in long.variable
    ]
    long.Structure = [
        replace(x, r"Length|Width" => "")
        for x in long.variable
    ]
    select!(long, Not(:variable))
    unstack(long, :Dimension, :value)
end
```

Here, we split the original column names into the flower structural elements and the measured dimension to reshape our long-format DataFrame into a wider one. Instead of using the indexing syntax and the bang to create the new columns, we used the property access syntax. For example, we used long.Dimension instead of long[!, :Dimension]. We generated the :Dimension and :Structure columns from the variable column that the stack function created by replacing the unwanted information with nothing. We took advantage of the fact that the replace function can take a regular expression that was created using the r"..." r string macro.

Once we had the new columns, we deleted the variable column, which is no longer needed, by using the mutating `select!` function. Again, we took advantage of the `Not` object to select all the columns except for `:variable`.

The `unstack` function takes a DataFrame in a long format and creates one in a wide format. In this case, we used two positional arguments after the input DataFrame. The first indicates the column that defines the name of the new columns, while the second is the column that contains the values that will fill the new columns. By default, `unstack` considers all the remaining columns as row identifiers. In this example, we distributed the values in the `:value` column into two new columns determined by the names in the `:Dimension` column: `Length` and `Width`.

Here, we learned how to read a DataFrame from a CSV file and perform everyday operations on that data. In particular, we learned how to describe, transform, and reshape our data. We also learned how to apply the *split-apply-combine* strategy using DataFrames thanks to the `groupby` and `combine` functions. Here, `groupby` splits a dataset into groups and `combine` applies functions to these groups to combine the results later. We used the `combine` function while using the `DataFrames` specification to apply functions to columns. `combine` can also take the function to apply to each group as the first argument. Therefore, we can use the `do` syntax to create an anonymous function that `combine` will apply to each subset.

The `select` and `transform` functions and their mutating versions, `select!` and `transform!`, are pretty useful for changing data columns. The `select` function will return a DataFrame that matches the rows in the input DataFrame but will only contain the specified columns, while `transform` will return a `DataFrame` with all the input columns plus the new ones. You can also apply those functions to the `GroupedDataFrame` object that `groupby` creates and, like `combine`, they can also use the *DataFrames specification*. The full specification consists of three elements:

- A column selector that determines the input columns. For example, you can select single columns using `Symbol`, `String`, or `Int`.

- A named or anonymous function that will be applied to the input columns to create new ones.

- The names for the output columns. You can use `Symbol`, `String`, or a vector of those objects.

These components are optional. For instance, if you omit the names for the output columns, `DataFrames` will automatically create column names, while if you omit the function, `DataFrames` will rename the columns.

There are other operations that the `DataFrames` package makes convenient but that we haven't discussed. Among those, we can name *sorting* and *database-style joining*. You can use the `sort` and `sort!` functions to sort the rows of a DataFrame. For joining, you can use an extended family of functions, such as `innerjoin` and `outerjoin`.

What's more, Julia has first-class support for missing values, and the `DataFrames` package offers a series of functions for working with tables containing missing data. Julia represents missing data using a missing object of the `Missing` type. Among the available `DataFrames` functions to work with this type, you will find `dropmissing` helpful for deleting all the rows with missing values.

With that, we've learned how to work with DataFrames in Julia. This data structure is critical for storing and transforming the data for further plotting through frameworks based on *The Grammar of Graphics*. Before we look at the plotting library, let's discuss data types and how they are represented visually.

Encoding data

As columns represent data variables in tidy DataFrames, with each column usually containing objects of one data type if we do not take `Missing` into account. In Julia, we can use different data types, such as the ones described in *Chapter 1, An Introduction to Julia for Data Visualization and Analysis*, to represent data. We can think of three main kinds of variables that are interesting for visualization within a *Grammar of Graphics* framework and that we can easily represent in Julia:

- **Categorical variables** describe a set of unordered categories or groups; for example, in the Iris dataset, the `Species` column is categorical. In Julia, we usually represent categorical values using objects of the `String` or `Symbol` type. `Bool` is also a common choice for representing **binary** or **dichotomous variables**, a categorical variable with only two levels. We can also take advantage of the `CategoricalArray` type from the `CategoricalArrays` package to represent this data compactly. The `DataFrames` package offers the `categorical!` function, which can convert a column into `CategoricalArray`. For example, `categorical!(iris, :Species)` will make the `iris` DataFrame use `CategoricalArray` for the `Species` column.

- **Ordinal variables** are like categorical variables, except that the group order is meaningful. Despite being ordered, we cannot know the actual distance between those groups. A classic example of an ordinal variable is shirt sizes; small, medium, or large. We can sort the shirt sizes; for instance, we know that a medium shirt is larger than a small one. However, we do not know how much larger the medium shirt is than the small one. We can use integer values, letters, or, even better, `CategoricalArray` with ordered levels to represent ordinal variables.

- **Quantitative variables** are numerical, and they usually come from a measurement. For example, the Iris dataset's length and width measures of sepals and petals are quantitative variables. They can be discrete, traditionally represented with integer numbers, or continuous, represented with floating-point numbers.

Then, we can classify these tree variable types into categorical and ordered variables, with the latter containing both ordinal and quantitative variables. We usually use different aesthetics or channels for those two groups. We map categorical variables to aesthetics, which allows us to differentiate between the categories, such as color hue and shape. Instead, we map ordered variables to aesthetics that imply the values' order, such as position, length, size, and color luminance and saturation.

In this section, we focused on the representation of tabular data and their variables. Therefore, we have omitted relational data, which we will analyze in *Chapter 7*, *Visualizing Graphs*, and spatial data, which we will analyze in *Chapter 8*, *Visualizing Geographically Distributed Data*. In the next section, we will learn how to plot tidy data using `Gadfly`.

Exploring data with Gadfly

`Gadfly` is a Julia implementation of A *Layered Grammar of Graphics*. However, some aspects will change from the grammar described initially for the R language. In this section, we will learn how to write the different grammar components using Julia and `Gadfly`. As with other grammar implementations, `Gadfly` makes extensive use of *defaults* to determine the unspecified components. This will allow us to start with a simple plot and build a more complex specification. Let's explore `Gadfly` and its syntax by plotting the data in the Iris dataset:

1. Execute the following code in the first cell:

```
begin
    import Pkg
    Pkg.activate(mktempdir())
    Pkg.add([
        Pkg.PackageSpec(name="RDatasets",
```

```
            version="0.7.7"),
        Pkg.PackageSpec(name="Gadfly",
            version="1.3.4"),
    ])
    using RDatasets, Gadfly
end
```

We have used Pluto's **Pkg cell** pattern to install specific versions of the required packages. We obtained this code with the help of the *pkghelper website* at `https://fonsp.com/article-test-3/pkghelper.html`. It imports the `Pkg` module, then uses the `activate` function to create a temporal environment. After that, it uses the `add` function to install the packages in the environment. The `add` function takes a list of `PackageSpec` objects, specifying the package names and versions. Finally, it loads the required packages. If you want to understand this process more deeply, please refer to the *Managing environments* section of *Chapter 1, An Introduction to Julia for Data Visualization and Analysis*.

Here, we have installed `Gadfly` 1.3.4 and `RDatasets` 0.7.7 for reproducibility – we wrote the examples with those versions. However, you can try the newest versions of those packages. For that, you can run `using RDatasets, Gadfly` instead of the *Pkg cell*.

2. In a new cell, execute `iris = dataset("datasets", "iris")`.

3. Run the following code block in a new cell:

   ```
   plot(iris, x="PetalLength", y="PetalWidth")
   ```

   As we can see, this code creates a scatter plot with the `PetalLength` variable on the *x* axis and `PetalWidth` on the *y* axis. Gadfly's `plot` function can take the data source as the first positional argument – in this example, the `iris` DataFrame. Then, the keyword arguments of the plot functions determine the *mappings* between data variables, column names, and *aesthetics*. For instance, in this plot specification, we have mapped the values in the `PetalLength` column to the x aesthetics. We have not specified the geometric object to use, but `Gadfly` has used points by default, creating a scatter plot.

4. Execute the following code in a new cell:

   ```
   plot(iris, x="PetalLength", y="PetalWidth", Geom.point)
   ```

   The preceding code creates a plot that is identical to the one created in the previous step. However, in this specification, we have explicitly set the *geometry* to `point` by passing Geom.point as a positional argument. Gadfly uses all positional arguments after the first one to indicate grammar components that are not data or mappings. In this case, we declared geometry from Gadfly's Geom module.

5. In a new cell, execute the following code:

```
categorical = plot(iris,
      x="PetalLength", y="PetalWidth",
      color="Species",
      Geom.point)
```

In this specification, we have mapped the Species categorical variable to the color aesthetics of the point geometry. As the variable is categorical, Gadfly chooses a set of distinguishable colors. Here, we have assigned the resulting plot object to a variable we will use in *Step 7*. The left panel of *Figure 5.2* shows the generated figure.

6. Execute the following code in a new Pluto cell:

```
ordered = plot(iris,
      x="PetalLength", y="PetalWidth",
      color="SepalLength",
      Geom.point)
```

In this case, we have mapped the color aesthetics of the point geometry to the SepalLength order variable. Therefore, we are getting the sepal length values mapped into a continuous color scale. The right panel of *Figure 5.2* shows the generated plot.

7. Execute hstack(categorical, ordered) in a new cell. Here, we are taking advantage of having the previous plots stored in variables to create a new figure containing both. The hstack function arranges the plots horizontally in two columns. Gadfly also offers the vstack and gridstack functions for other layouts. The following figure shows the resulting plot:

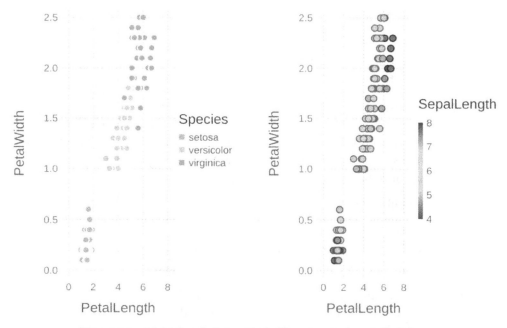

Figure 5.2 – Multiple subplots with Gadfly using the hstack function

8. In a new cell, run the following code:

```
plot(iris,
    x="PetalLength", y="PetalWidth", color="Species",
    Geom.point, Geom.line)
```

The preceding code takes two geometries instead of one: point and line. In Gadfly, each Geom that is explicitly outside a layer creates its own *layer*. So, this plot has two layers that share the same mappings. For example, both the line and point geometries share the same color aesthetics.

9. Execute the following code block in a new cell:

```
plot(iris, x="PetalLength", y="PetalWidth",
    layer(color="Species", Geom.point), Geom.line)
```

Here, we have explicitly created a *layer* for the point geometry to only map the point color to the Species value without affecting the line's color. We have used the layer function, which can take geometries and mappings as the plot function.

10. Execute the following code in a new cell:

```
plot(iris,
x="PetalLength", y="PetalWidth", color="Species",
Stat.smooth)
```

This specification applied a smooth statistical transformation to the input data before plotting. Gadfly offers a series of statistics in the Stat module that you can pass as positional arguments to the plot and layer functions. The generated plot is probably not what we want, so let's improve this by using two layers – one to show the data points and the other for smooth but using lines.

11. Run the following in a new Pluto cell:

```
plot(iris,
x="PetalLength", y="PetalWidth", color="Species",
layer(Stat.smooth, Geom.line), Geom.point)
```

This plot specification creates two layers – one implicitly for Geom.point and the other explicitly for plotting the smooth line using the layer function. Note that Gadfly, by default, draws the layers while following the plot arguments in reverse order. In this case, Gadfly plots the layer with the points as they are the last positional argument. The following figure shows the plot we have created in this step:

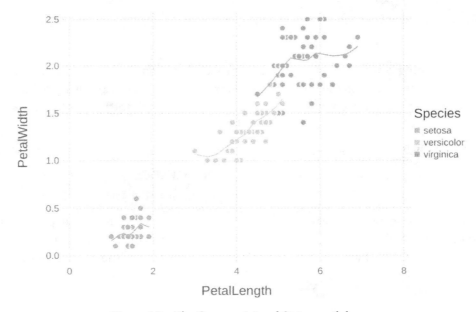

Figure 5.3 – The Geom.point and Stat.smooth layers

12. Execute `plot(iris, x="PetalLength", y="PetalWidth", size="SepalLength", Geom.point)` to map the sepal lengths to the point size aesthetics. We will see that the size of the points is too large for us to get any information. Let's fix that by adjusting the size scale.

13. In a new cell, execute `min_sl, max_sl = extrema(iris.SepalLength)` to store the minimum and maximum lengths of the Iris sepals.

14. Execute the following code in a new cell:

```
sizemap(p::Float64; min=1mm, max=2.5mm) = min + p*(max-min)
```

The preceding code creates a function that will take the variable value as a proportion and return the radius in mm.

15. Execute the following code block in a new cell:

```
plot(iris,
    x="PetalLength", y="PetalWidth",
    size="SepalLength",
    Geom.point,
    Scale.size_radius(sizemap,
    minvalue=min_sl, maxvalue=max_sl))
```

Here, we used the `size_radius` scale from the `Scale` module to fix the default *scale* size of our plot. We determined the minimum and maximum value of `SepalLength` to allow `Gadfly` to calculate a proportion for the `sizemap` function defined in the previous step. Now, our `SepalLength` variable is correctly mapped to the point `size` aesthetics. Note that scales are shared across layers, so layers cannot define their scales. By running the code in this step, you will see the plot shown in the following figure:

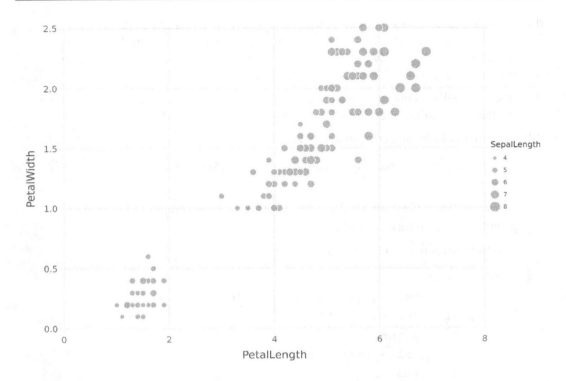

Figure 5.4 – Gadfly plot mapping a third variable to the point radius

In the preceding code, we mapped three data variables using the size and x and y position aesthetics for the point geometry. The preceding figure shows that `Gadfly` automatically creates a key for the variables that we haven't mapped to spatial coordinates.

16. In a new cell, execute the following code:

```
plot(iris,
     x="PetalLength", y="PetalWidth",
     ygroup="Species",
     Geom.subplot_grid(Geom.point))
```

The preceding code creates a *small multiple* or *trellis plot*. We can specify a *facet* element in `Gadfly` by calling the `subplot_grid` type from the `Geom` module and defining the `xgroup` and `ygroup` aesthetics. The following figure shows the facet plot we have created:

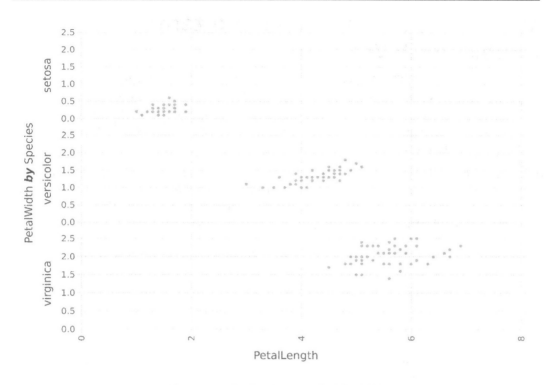

Figure 5.5 – Trellis plot created with Gadfly

Through these steps, we have quickly explored how to represent the different components of the grammar in Gadfly. The preceding figure also maps three variables. We map two of them to the point geometry's *x* and *y* coordinate aesthetics. We used the third variable, which is categorical, to divide the data into groups, thus creating multiple plots that share the same mappings.

This section has taught us how to write a plot specification using the Gadfly implementation of A *Layered Grammar of Graphics*. In the next section, we will look at Vega-Lite's extension of the grammar for including interactive plots.

Introducing the Grammar of Interactive Graphics with VegaLite

Vega-Lite is a JavaScript library that extended the *Grammar of Graphics* to a *Grammar of Interactive Graphics*. It uses *Vega* as a backend, and you need to write the plot specifications in *JSON*. Its grammar refers to *mark* as *geometry*, *channel* as *aesthetic*, and *encoding* as *mapping*. The grammar provides statistics such as density, inside transformations, and interactivity mainly through selection.

You can use Vega-Lite from Julia thanks to the `VegaLite` package. Its `@vlplot` macro allows you to write the specification in Julia. In the `@vlplot` macro, the first positional argument indicates the mark to use, and the keyword arguments indicate channels and encodings. Let's look at some examples of the syntax in action:

1. Create a new Pluto notebook and execute the following code in the first cell:

    ```
    begin
        using Pkg
        Pkg.activate(temp=true)
        Pkg.add([
            PackageSpec(name="VegaLite", rev="50be67d"),
            PackageSpec(name="RDatasets")
        ])
        using VegaLite, RDatasets
    end
    ```

    This cell is similar to the **Pkg** cell we used in the previous section. However, there are some crucial differences. First, we have indicated a Git commit using the `rev`, for Git revision, keyword argument. We did this since the feature we want for `VegaLite` is not included in a tagged release yet. Then, we added `RDatasets` without indicating the desired version so that Julia will install the latest version possible.

2. In a new Pluto cell, run `iris = dataset("datasets", "iris")` to load the Iris dataset.

3. Run the following code in a new cell:

    ```
    iris |> @vlplot(:circle,
        x="PetalWidth", y="PetalLength",
        size="SepalLength")
    ```

Here, we passed the data using the Julia pipe operator, | >, to the @vlpot macro. The first positional argument indicates the mark; in this case, we used the :circle mark, but you can also see a similar plot by using :point or :square. We also used a Symbol to indicate the mark, but you can use a String if you prefer; both are interchangeable in the @vlpolt syntax. Then, we used keyword arguments to encode the PetalWidth variable to the *x* position channel, PetalLength to the *y* position, and SepalLength to the size of the circle mark. Note that we haven't optimized the scale of the circle sizes in this case. For instance, as shown in the following figure, the number 2 appears on the key for SepalLength, even when there is no sepal shorter than 4.3 cm in the dataset:

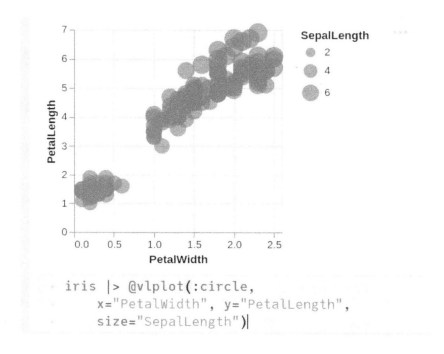

```
iris |> @vlplot(:circle,
    x="PetalWidth", y="PetalLength",
    size="SepalLength")|
```

Figure 5.6 – VegaLite plot mapping SepalLength to circle size with the automatic scale

4. Run min_sl, max_sl = extrema(iris.SepalLength) to get the minimum and maximum values for the sepal length.

5. In a new cell, run the following code block:

```
iris |> @vlplot(:circle,
x="PetalWidth", y="PetalLength",
size={"SepalLength",
        scale={domain=[min_sl, max_sl]}})
```

Note that we have corrected the scale for the circle size property channel by explicitly indicating its domain. The following figure shows the resulting plot:

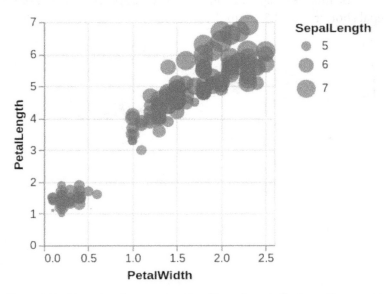

Figure 5.7 – VegaLite plot mapping SepalLength to circle size with our custom scale

Here, we have encoded a third variable, SepalLength, using the circle's size. Note that Vega-Lite uses the pixel area for the size properties of point, circle, and square. Therefore, even though this figure looks similar to *Figure 5.4*, it uses area rather than radius to encode SepalLength.

6. In a new cell, run the following line of code:

```
iris |> @vlplot(:circle,
x="PetalWidth", y="PetalLength",
tooltip="SepalLength")
```

Here, we have encoded SepalLength to the circle tooltips rather than to their size. That's because, in Vega-Lite, tooltip is just another encoding. Now, you can move your mouse over the circles to see the sepal length for a data point.

7. Execute the following code block in a new cell:

```
iris |> @vlplot(:circle,
    x="PetalWidth", y="PetalLength",
    selection={brush={type="interval"}},
    color={condition={selection="brush",
        value="red"}, value="lightgrey"})
```

The preceding code creates an interactive plot thanks to the use of selection. Vega-Lite defines two selection types: interval, the one we are using here, and point. To select an interval or region, you need to drag on the plot by pressing and holding down the left mouse button while moving your mouse. Note that we are naming our interval selection brush so that the brush variable will keep track of the selected points. In this case, Vega-Lite will color the selected points red and the others light gray. condition on the color encoding indicates this behavior. Let's use this coloring brush in a more exciting example.

8. Execute the following code in a new cell:

```
iris |> @vlplot(repeat={
    column=["PetalLength", "PetalWidth"],
    row=["SepalLength", "SepalWidth"]}) +
    @vlplot(:circle,
        x={field={repeat="column"},
            type="quantitative"},
        y={field={repeat="row"},
            type="quantitative"},
        selection={brush={type="interval"}},
            color={condition={selection="brush",
                value="red"}, value="lightgrey"},
        width=250, height=200)
```

Here, we defined two @vlplot specifications. The first one uses repeat to create four scatter plots; the first column will encode PetalLength on the *x* position channel while the second column will use PetalWidth instead. Then, the first row encodes SepalLength on the *y* position channel, and the second row encodes SepalWidth instead.

The four plots will use the second `@vlplot` specification. This specification is similar to the one in the previous step but with some minor differences. When mapping the *x* and *y* position channels, we indicate that the plot should take the variable names from the column and row specified repeatedly. In this case, we also explicitly pointed out that the variables are quantitative. Then, we fixed the plot size by setting `width` and `height`. Note that if you select a series of points in one of the four plots, Vega-Lite will highlight the same observations on the other three plots, as shown in the following figure:

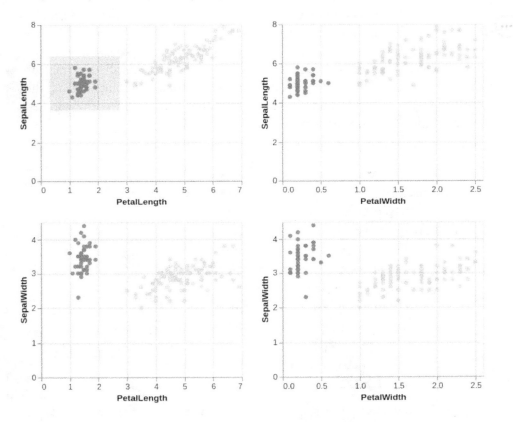

Figure 5.8 – VegaLite interactive selection

This section quickly explored some aspects of the `VegaLite` syntax for the *Grammar of Interactive Graphics* that you will need to get started with the package. In the next section, we will learn about the basics of Makie's *Algebra of Graphics*.

Makie's Algebra of Graphics

The `AlgebraOfGraphics` package is another Julia implementation of the *Grammar of Graphics*. It renders the plots using Makie. Therefore, you need to load a Makie backend, such as `CairoMakie`, to work with it. The grammar of the *Algebra of Graphics* has the following main components:

- **Data**: The `data` function indicates the dataset to be used.

- **Mapping**: `AlgebraOfGraphics` uses the `mapping` function to indicate the *mapping* between data variables and aesthetics. The first three positional arguments designate the variables that have been mapped to the *x*, *y*, and *z* positions. The mapping function maps other aesthetics, such as `color` and `markersize`, using keyword arguments. The function also implements a *pair syntax* for *transformations* of the data.

- **Visual**: This is related to *geometry* on other grammar implementations as it designates what type of named plot we are creating. `AlgebraOfGraphics` uses the `visual` function to indicate a Makie named plot using *CamelCase*; for example, `Scatter` or `BoxPlot`.

- **Analyses**: These are the *statistics* of this grammar.

In the *Algebra of Graphics*, `Layer` is the main building block, indicating a plot specification. We can create a *layer* by combining the components listed previously. Then, the *Algebra of Graphics* allows you to perform algebraic operations on the specifications. We can merge specifications using the `*` operator and superimpose layers using the `+` operator. Being able to cleanly separate the components and the algebraic rules to combine them gives a lot of flexibility and composability to the *Algebra of Graphics*.

Let's look at some of the parts of this grammar in action. To do so, we need to create a new Pluto notebook and run the following code in the first cell:

```
begin
using RDatasets, AlgebraOfGraphics, CairoMakie
    set_aog_theme!()
    iris = dataset("datasets", "iris")
    plt = data(iris) *
        mapping(:PetalWidth, :PetalLength,
            markersize=:SepalLength) *
        visual(Scatter)
    draw(plt)
end
```

The preceding code block creates a scatter plot similar to the ones shown in *Figure 5.4* and *Figure 5.7*. We have mapped `PetalWidth` and `PetalLength` to the *x* and *y* positions, respectively, and `SepalLength` to `markersize` using the `mapping` function. Note that we used the `data` function to indicate the input dataset and used the `*` operator to combine it with the objects returned by `mapping` and `visual`. In this case, we are using `Scatter` as the *visual* for our plot. Finally, we can create a plot from the specification by calling the `draw` function. The plot uses the theme of `AlgebraOfGraphics`, as set by the `set_aog_theme!` function.

Summary

In this chapter, we learned about the *Grammar of Graphics* and its three main Julia implementations. We now understand the underlying idea behind the grammar while knowing what the main syntax and nomenclature variations in the Julia ecosystem are. We also gained some skills for handling data in Julia using the `DataFrames` package, as data is one of the main components of grammar. Then, we explored `Gadfly` and learned how to take advantage of its sensitive use of defaults to create complex visualizations with a few lines of code. After that, we learned how to create interactive plots using `VegaLite` and its *Grammar of Interactive Graphics*. Finally, we explored the `AlgebraOfGraphics` composable.

In the next chapter, we will take advantage of these libraries and others to create different statistical plots in Julia.

Further reading

The Grammar of Graphics is an extensive topic; you can learn more about the different variations of grammar by reading the following articles:

- *Wickham, Hadley. A layered grammar of graphics. Journal of Computational and Graphical Statistics 19.1 (2010): 3-28.* https://doi.org/10.1198/jcgs.2009.07098

- *Satyanarayan, Arvind, et al. Vega-lite: A grammar of interactive graphics. IEEE transactions on visualization and computer graphics 23.1 (2016): 341-350.* https://www.doi.org/10.1109/TVCG.2016.2599030

Munzner, Tamara. Visualization analysis and design. CRC Press, 2014, can help you learn how to choose the best marks and channels for your visualizations based on your variables.

For more information about the Julia implementations, we strongly recommend reading the documentation of your package of interest:

- Gadfly: http://gadflyjl.org/stable/

- AlgebraOfGraphics: http://juliaplots.org/AlgebraOfGraphics.jl/stable/

- VegaLite: https://www.queryverse.org/VegaLite.jl/stable/

If you are interested in VegaLite, reading the *Vega-Lite* documentation is also recommended; you can find it at https://vega.github.io/vega-lite/docs/.

6
Creating Statistical Plots

Creating statistical plots is a standard *data analysis* task, especially during data exploration. It is an essential part of *data visualization*, helping make meaningful visual representations for our data. It is crucial, as in many cases, that we learn more from our data by looking at it than by exclusively analyzing its summary statistics. *Anscombe's quartet* is an example of this as its four datasets show similar descriptive statistics but different distributions we can see after plotting them. *Figure 6.1* shows these datasets with a *Pearson correlation coefficient*, *r*, of *0.82*, but various joint distributions.

Also, we can rely on statistical plots to effectively *communicate* our findings to the world – a common *data visualization* task. Some visualizations, such as histograms, are easily understood by people from many backgrounds. Others, such as boxplots, are better suited for a statistically versed audience:

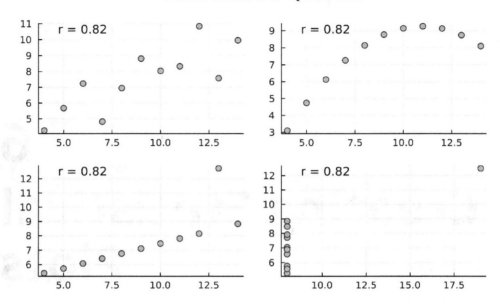

Figure 6.1 – Anscombe's quartet

In this chapter, we will learn how to create statistical plots using Julia. To that end, we will take advantage of the packages we learned about in *Chapter 5, Introducing the Grammar of Graphics*. We will also explore the `StatsPlots` package, which offers statistical plots and `DataFrames` support for `Plots`.

In particular, we will learn how to visualize univariate and bivariate distributions through this chapter. We will also learn how to visualize clustering results and compare distributions between clusters. Those skills will be advantageous to you when visually analyzing or sharing data.

In this chapter, we're going to cover the following main topics:

- Introducing `StatsPlots`
- Visualizing univariate distributions
- Plotting bivariate distributions and regressions
- Drawing regression lines
- Creating marginal plots
- Visualizing clustering results
- Comparing between groups

Technical requirements

For this chapter, you will need Julia, `Pluto`, and a web browser with an internet connection. For the plot examples in this chapter, we have used the following packages and versions:

- `Plots 1.31.1`
- `StatsPlots 0.14.30`
- `CairoMakie 0.6.6`
- `AlgebraOfGraphics 0.6.0`
- `Gadfly 1.3.4`
- `RDatasets 0.7.6`
- `Distributions 0.25.37`
- `Clustering 0.14.2`
- `Distances 0.10.7`
- `PlutoUI 0.7.23`

If you get errors when executing this chapter's code examples, they will probably come from using different package versions. In that case, you can install the version listed previously while following the indications in the *Managing environments* section of *Chapter 1, An Introduction to Julia for Data Visualization and Analysis*. If you are working with `Pluto`, as suggested in many sections of this chapter, you can use the *Pkg cell* pattern shown in the *Exploring data with Gadfly* section of *Chapter 5, Introducing the Grammar of Graphics*. Alternatively, you can run the example Pluto notebooks stored in the `Chapter06` folder of this book's GitHub repository at `https://github.com/PacktPublishing/Interactive-Visualization-and-Plotting-with-Julia`. Pluto notebooks hold their environment, so you should not have problems with the package versions. You can find the Pluto notebook files and their HTML versions, along with stored outputs, in that folder.

Introducing StatsPlots

Before describing different statistical plots, let's introduce the main Julia package we will use in this chapter: the `StatsPlots` package. It defines plotting recipes to create statistical plots using the `Plots` package. It also adds the `@df` macro to support the `DataFrames` package. Furthermore, the `StatsPlots` package offers type recipes for some of the types defined on packages from the *JuliaStats organization*. Among those, we can find the `Clustering`, `Distributions`, and `MultivariateStats` packages. For example, we can call the `plot` function on a `Cauchy` object from the `Distributions` package to plot our *Cauchy distribution*. We will learn more about type recipes in *Chapter 14, Designing Your Own Plots – Plot Recipes*. Let's explore the syntax and basic features of the `StatsPlots` package using the *Iris dataset* and Pluto:

1. Create a new Pluto notebook and execute `using RDatasets` in the first cell.

2. In a new cell, execute `iris = dataset("datasets", "iris")` to load the Iris dataset.

3. Execute the following code line in a new cell:

   ```
   using StatsPlots
   ```

   The preceding code loads the `StatsPlots` package, which re-exports the `Plots` package. So, by loading `StatsPlots`, you gain access to all the functions that have been exported in the `Plots` package. Therefore, there is no need to explicitly load `Plots`. `StatsPlots`, similar to `Plots`, will use the *GR backend* by default.

4. Execute `plotly()` in a new cell to change the backend from *GR* to *Plotly*. We can ignore the message telling us to load `PlotlyBase` to export the figures as we do not plan to do that.

5. Execute the following code in a new cell:

   ```
   @df iris scatter(:PetalLength, :PetalWidth)
   ```

   The preceding code creates a scatter plot with `PetalLength` on the *x* axis and `PetalWidth` on the *y* axis. Note that we have used the `@df` macro to indicate that we are taking our *data* from the `iris` DataFrame, and we have stated the columns to use. Note that we need to pass each *column* name using a `Symbol` rather than a `String`. The code we executed in this step is equivalent to running `scatter(iris.PetalLength, iris.PetalWidth)`, which explicitly passes the columns to the `scatter` function. Throughout this chapter, we will use the `@df` macro to indicate the data we want to plot. Therefore, let's explore it in more depth.

6. In a new cell, execute the following code:

   ```
   @df iris scatter(pi .* :PetalLength .* :PetalWidth)
   ```

   Here, we plotted a point for each flower; the *x* axis indicates the row number, while the *y* axis shows an estimation of the petal area. As we pass a single positional argument to the plotting function, it plots its values on the *y* axis and automatically assigns the indexes to the *x* axis. Here, the `@df` macro allows us to perform *operations* on the columns; in this case, element-wise multiplication using the broadcasted multiplication operator, .*, to calculate the area. We used an ellipse to approximate the petal shape to estimate the area for this example.

7. Execute the following code line in a new cell:

```
@df iris scatter(cols(3:4))
```

The preceding code plots two series of points for each row in the iris DataFrame – one series for column number 3, PetalLength, and another for column number 4, PetalWidth. This code example uses cols to tell the @df macro which *columns* to select using a numeric range. We can also use cols to choose a single column using the column index or a variable containing the column name, as we will see in the following steps. If you want to select all the columns in a DataFrame, you can use cols() inside the call to the @df macro.

8. Execute using PlutoUI in a new cell.

9. Create a new cell and execute the following code:

```
@bind column_name Select(Symbol.(names(iris)[1:4]))
```

The preceding code defines the column_name variable, whose value is selected using a dropdown menu created with Select from the PlutoUI package. We used the names function to get a vector of strings with the column names. Then, we selected the first four names. Finally, we converted each String into a Symbol before passing the resulting vector to Select. This way, we have ensured that column_name will contain the selected column name as a Symbol, as needed for cols and the @df macro.

10. In a new cell, execute the following code block:

```
@df iris scatter(
    cols(column_name),
    ylabel=column_name,
    xlabel="Row number",
    legend=false)
```

The preceding code creates a scatterplot with the selected variable on the *y* axis and the row number on the *x* axis. Here, we used cols to tell the @df macro the column to choose from a variable that contains Symbol, which indicates the column's name. For this plot, we are also using the column_name variable to define the label for the *y* axis accordingly. Thanks to Pluto reactivity, you can select another column in the dropdown menu and see the updated plot. The following figure shows the resulting plot:

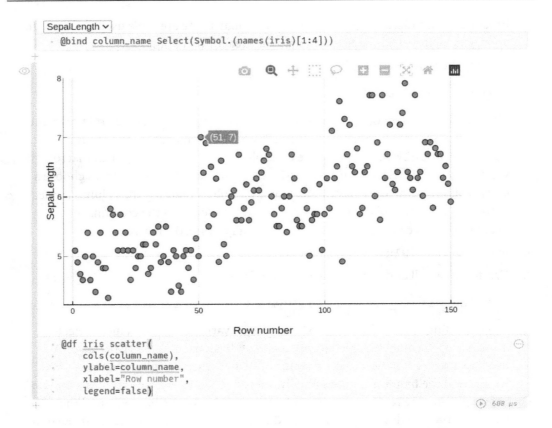

Figure 6.2 – Interactive scatter plot created with StatsPlots, Plotly, and Pluto

By completing these steps, we have learned about the syntax for plotting tabular data with `StatsPlots` and the `@df` macro. Also, in the last few steps, we used Pluto to create a visualization for which we can easily change the input data through `PlutoUI` widgets. That will come in handy when analyzing multiple variables. Also, the Pluto approach is very flexible as it gives you complete control over the plot specification to use. However, you will find it convenient to use the `StatsPlots` dashboard for a quick visual exploration of your data.

The `StatsPlots` dashboard is an application that's created with the `dataviewer` function, which offers an interactive GUI for producing standard plots. It uses `Interact` to define the interactive elements of the interface. And because the application uses `Interact`, we also need to load an extra package to deploy it. In particular, you can deploy the application using `Mux`, `Blink`, or Jupyter. For the last one, you can follow the setup instructions in the *Setting up Jupyter for Interact* section of *Chapter 3, Getting Interactive Plots with Julia*. You will also find an example of how to use `Interact` and `Mux` in that chapter. So, let's see an example using `Blink`:

1. Open a Julia terminal and execute the following code block:

    ```
    import Pkg
    Pkg.activate(temp=true)
    Pkg.add([Pkg.PackageSpec(name="StatsPlots",
            version="0.14.30"),
        Pkg.PackageSpec(name="Interact", version="0.10"),
        Pkg.PackageSpec(name="Blink", version="0.12"),
        Pkg.PackageSpec(name="RDatasets",
            version= "0.7.6")])
    ```

    To ensure the reproducibility of this example, we have installed particular versions of the required packages in a temporal environment.

2. Execute the following code block:

    ```
    using RDatasets
    iris = dataset("datasets", "iris")
    ```

    The preceding loads the Iris dataset that we will use for this example.

3. Execute `using StatsPlots, Interact, Blink` to load the libraries needed to use the dashboard and deploy it.

4. Run the following code in the Julia terminal:

    ```
    window = Window()
    body!(window, dataviewer(iris))
    ```

The preceding code creates an *Electron* window and fills it using the StatsPlots `dataviewer` function. The `dataviewer` function creates the dashboard for the data given as input. You will see a dashboard similar to the following:

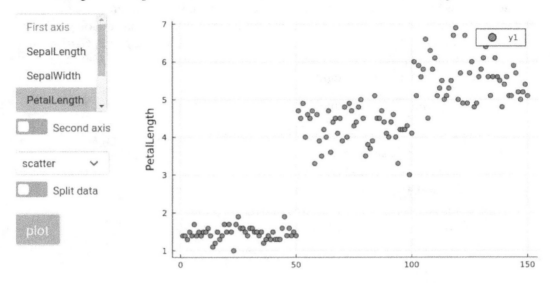

Figure 6.3 – The StatsPlots dashboard

Now, we can make the most of the `StatsPlots` package; we will use it throughout this chapter to plot different statistical plots. We will start by visualizing single variable distributions.

Visualizing univariate distributions

Visualizing the distribution of a single random variable is a pretty common task. Therefore, there are multiple named plots we can use for this. These visualizations help us see a variable distribution's *location*, *dispersion*, and *shape*. They are also helpful in *comparing* the distribution of different variables or the same variable in different conditions. This section will discuss the most common plots for visualizing univariate data and show you how to create them with Julia while focusing on `StatsPlots`.

In this section, we will look at examples for `StatsPlots`, `Gadfly`, and *Makie*. Note that those packages can export functions with the same name; for instance, `StatsPlots` and `AlgebraOfGraphics` export `histogram`. If you want to load multiple plotting packages in the same session, you will need to use the *qualified names* for those functions, such as `StatsPlots.histogram` or `AlgebraOfGraphics.histogram`; otherwise, you will see an `UndefVarError` message. For this section, we recommend testing each plotting package on its own `Pluto` notebook to avoid errors. At the beginning of each notebook, you can include a **Pkg cell** to install the required version of each package. For a notebook using `StatsPlots`, you should run the following code in the first cell:

```
begin
    import Pkg
    Pkg.activate(temp=true)
    Pkg.add([Pkg.PackageSpec(name="StatsPlots",
        version="0.14.30"),
        Pkg.PackageSpec(name="RDatasets",
            version="0.7.6")])
    using RDatasets, StatsPlots
    plotly()
end
```

The final expression loads the *Plotly* backend to take advantage of its interactivity. This code also loads `RDatasets` to allow us to load the Iris dataset later. It also imports `Pkg` so that we can call `Pkg.add` in other cells if we need it, but we can only have a single call to `activate` per notebook.

To use *Makie*, we will count on its `CairoMakie` backend for this chapter. With *Makie*, we will also explore the syntax of the `AlgebraOfGraphics` package to create some of the plots. Therefore, to work with *Makie* through this chapter, you should execute the following code at the beginning of the Pluto notebook:

```
begin
    import Pkg
    Pkg.activate(temp=true)
    Pkg.add([Pkg.PackageSpec(name="AlgebraOfGraphics",
            version="0.6.0"),
        Pkg.PackageSpec(name="CairoMakie",
            version="0.6.6"),
        Pkg.PackageSpec(name="RDatasets",
            version="0.7.6")])
    using RDatasets, AlgebraOfGraphics, CairoMakie
    set_aog_theme!()
end
```

Finally, to load `Gadfly`, you can execute the following code in the first cell of the `Pluto` notebook:

```
begin
    import Pkg
    Pkg.activate(temp=true)
    Pkg.add([Pkg.PackageSpec(name="Gadfly",
        version="1.3.4"),
        Pkg.PackageSpec(name="RDatasets",
            version="0.7.6")])
    using RDatasets, Gadfly
end
```

This section's examples will use the data in the Iris dataset, which we can load by using the `RDatasets` package and executing the following code line:

```
iris = dataset("datasets", "iris")
```

In particular, we will use *petal length* and *sepal width* as examples of continuous variables for plotting in this section.

Let's start with one of the most famous plots to learn about variable distribution: the histogram.

Drawing histograms

A **histogram** offers a simple way to approximate the **probability density function** (**PDF**) of a random variable distribution. To create a histogram, we need to divide the variable domain into discrete bins. Then, we must count the number of observations in each bin. Because this is a pretty standard procedure, all the plotting packages can do this automatically. Plotting a histogram with `StatsPlots` is as simple as calling the `histogram` function:

```
@df iris histogram(:PetalLength)
```

To test this code, you must create a new `Pluto` notebook and execute the code proposed in this chapter's *Visualizing univariate distributions* section to load `StatsPlots` and the Iris dataset. Creating histograms is so popular that the `Plots` package already exports this `histogram` function. However, in the preceding code, we used `StatsPlots` to take advantage of the `@df` macro. The following figure shows the histogram that was created from that line of code:

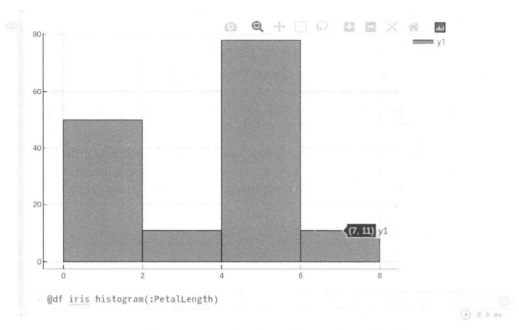

```
@df iris histogram(:PetalLength)
```

Figure 6.4 – Histogram created with StatsPlots and the Plotly backend

The preceding histogram has four bins; StatsPlots has automatically determined that bin number. Since the histogram's shape depends on the number and location of the bins, the histogram function has a bins keyword argument to change the criterion used to determine them. The different criteria use mathematical formulas that return floating-point numbers. Those numbers are subsequently rounded to the next integer value to determine the number of bins. StatsPlots also ensures that the number of bins will be an integer in the closed interval *[1, 10000]*. You can choose one of the following criteria by passing a Symbol to the bins argument:

- **Square-root choice**: You can select this by giving the :sqrt symbol to bins. It is simply the square root of the number of samples.

- **Sturges' formula**: You can choose this with :sturges. It calculates the number of bins using the logarithm in base 2 of the number of samples and adding one. It assumes an approximately normal distribution; therefore, it performs poorly with asymmetrical distributions. The optimal sample number to apply this rule lies between 30 and 200.

- **Rice rule**: This can be selected with :rice. It sets the number of bins to two times the cube root of the number of samples.

- **Scott's normal reference rule**: This can be selected using `:scott`. As its name suggests, it assumes normally distributed data. It uses the standard deviation to measure the dispersion of the values; therefore, it is sensitive to outliers.

- **Freedman-Diaconis rule**: You can choose this using `:fd` or `:auto`. It is the rule that `StatsPlots` uses by default. It is similar to Scott's rule, but it uses the interquartile range rather than standard deviation to measure data dispersion. Therefore, this rule is more robust to outliers in the data.

For example, you can run the following code block to compare the different criteria in your selected variable. You only need to have `StatsPlots` and the Iris dataset loaded in your Pluto notebook – note that loading another plotting library to export plots or histograms will make this code fail:

```
@df iris plot(
    histogram(:PetalLength, bins=:sqrt,
        title="Square-root choice"),
    histogram(:PetalLength, bins=:sturges,
        title="Sturges' formula"),
    histogram(:PetalLength, bins=:rice,
        title="Rice rule"),
    histogram(:PetalLength, bins=:scott,
        title="Scott's normal reference rule"),
    histogram(:PetalLength, bins=:fd,
        title="Freedman-Diaconis rule"),
    titlefontsize=10,
    legend=false)
```

Here, we used a single call to the `@df` macro to plot multiple histograms from the same dataset in that block. The following figure shows the plot that was created:

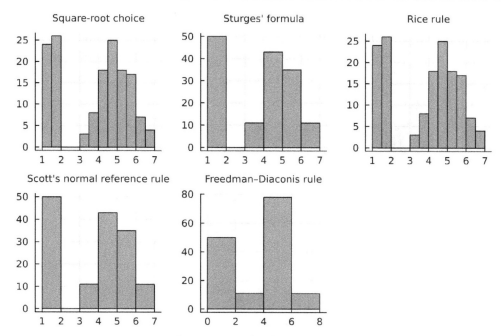

Figure 6.5 – Different criteria for choosing the number of bins

As you can see, we can get different feelings about the underlying distribution depending on the chosen criteria. Therefore, we recommend testing different data binning techniques to have a good sense of its probability distribution. In general, using too many bins increases the noise in the plot, especially when we have few data points. However, we should consider that using too few bins can hide important details. For example, in the preceding figure, using 12 bins, as selected by the square-root rule, highlights that the distribution has at least two modes. However, the separation between the two peaks is not noticeable when we use the four bins that the Freedman-Diaconis rule chooses. Later in this chapter, we will see that these separated peaks come from the presence of different species in the Iris dataset, making the petal length distribution multimodal. We must note that our multimodal distribution is not normal, breaking the assumptions of some of the rules.

Because the communicated distribution depends on data binning, the bins keyword argument of the histogram allows for finer binning control. In particular, you can pass a single integer to the bins argument to suggest the desired number of bins. Then, the algorithm will choose a bin number around that value. If you want to ensure a given number of bins, the best option is to provide a vector with the desired edges to the bins keyword argument. Each pair of consecutive numbers in the vector indicates the start and end values for the interval representing the bin. Those intervals are closed on the left and open on the right. We can take advantage of the range function to create this.

For example, the following code block will create two histograms when executed in the notebook with `StatsPlots` and the Iris dataset loaded. For the first histogram, we suggest using 10 bins, while for the second one, we ensure 10 bins by explicitly providing the bin boundaries with the `range` function:

```
@df iris plot(
    histogram(:PetalLength, bins=10,
        title="Integer"),
    histogram(:PetalLength,
        bins=range(
            minimum(:PetalLength),
            maximum(:PetalLength) + 0.1,
        length=11),
    title="Range"),
    legend=false)
```

Here, we created the vector of bin boundaries using the `@df` macro to operate on the column names. In particular, we created a range of numbers that go from our sample's minimum value to the largest one. As the intervals that determine the bins are open on the right, they do not contain the last value indicated. Therefore, to ensure that the `histogram` function counts the maximum value in the final interval, we have added a small quantity to that value, `0.1`. We used the `length` keyword argument of `range` to divide it. Note that since the vector elements indicate the limits of the intervals, we need a vector that is one element longer than the desired number of bins. Therefore, here, we have stated a length of `11` to get 10 bins. The following figure shows the resulting plot:

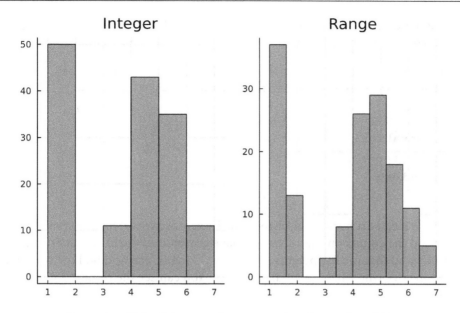

Figure 6.6 – Using integers and ranges to select the number of bins

The `bins` keyword argument is critical for drawing histograms. However, the `histogram` functions from `Plots` and `StatsPlots` also offer other keyword arguments so that you can customize your histograms further. The most important ones are as follows:

- `bar_width`: As its name suggests, this keyword argument determines the width of the histogram's bars.

- `orientation`: By default, this argument is set to `:vertical` to create the classical histograms. However, you can create a horizontal histogram with the variable in the *y* axis using `:horizontal` instead.

- `weights`: By default, all the observed data point weights are the same. We can change that by using this keyword argument to assign different weights to each data point.

- `normalize`: This keyword argument affects the height of the bars and, therefore, the values on the *y* axis. By default, there is no normalization, so the bar heights are identical to the counts, taking the weights into account. We should set this argument to `false` or `:none` to explicitly avoid normalization. Then, if we set `normalize` to `true` or `:pdf`, the bar heights are normalized so that the bins' total area is one. This is equivalent to having a discrete probability density function, formally known as a **probability mass function**. Setting `normalize` to `:probability` will make the sum of the bar heights, rather than their areas, 1.

The last option is to set this keyword argument to `:density`. In this case, the area of the bars, rather than the height, will be equivalent to the weighted counts.

As an example, in the following `StatsPlots` code, we have set some of these keyword arguments to values other than the defaults to see their impact:

```
@df iris histogram(:PetalLength,
    orientation=:horizontal,
    normalize=:probability,
    bar_width=1,
    legend=false)
```

In particular, we created a horizontal histogram. Therefore, we will have our variable on the *y* axis and the counts on the *x* axis. We have normalized the histogram using `:probability` so that the bar heights add up to 1. Finally, we modified the bar widths to 1 rather than cover the whole interval of the bin. You can see the resulting plot in the following figure:

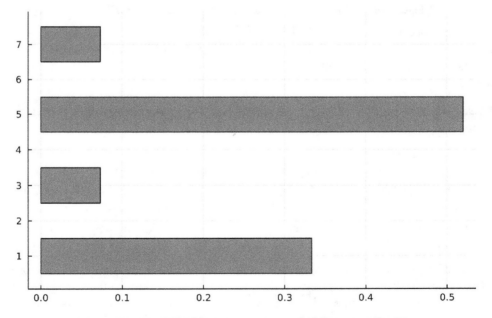

Figure 6.7 – Horizontal histogram using probability normalization

So far, this section has focused on the `histogram` function from the `Plots` and `StatsPlots` packages. However, multiple plotting packages offer a way to draw histograms; one is `Gadfly`. The `Gadfly` package offers the `Geom.histogram` geometry, a combination of `Geom.bar` and `Stat.histogram`, to easily create histograms. For instance, you can create a new `Pluto` notebook, load `Gadfly` and the Iris dataset, and execute the following code to generate a histogram of the petal length with `Gadfly`:

```
plot(iris, x=:PetalLength, Geom.histogram)
```

Finally, you can also create histograms using the *Makie ecosystem* – to test the code, create a new `Pluto` notebook while loading `CairoMakie`, `AlgebraOfGraphics`, and the Iris dataset. Makie exports the `hist` function to draw histograms; for example, the histogram of petal lengths will be as follows:

```
hist(iris.PetalLength)
```

We can also use the `AlgebraOfGraphics` package to create a histogram using the same function. We need to indicate the Makie function to the `visual` function using *CamelCase*. So, for example, the `hist` function becomes `Hist`:

```
data(iris) * mapping(:PetalLength) * visual(Hist) |> draw
```

The `hist` function and the `Hist` visual allow you to set the number of bins with the `bins` keyword argument and normalize the histogram with `normalization`. The `AlgebraOfGraphics` package also offers the `histogram` function, which supports the `weights` keyword argument. The following code creates the histogram using the `histogram` function:

```
data(iris) * mapping(:PetalLength) * histogram() |> draw
```

The histograms we have seen in this section estimate the underlying PDF of a variable distribution by discretizing the variable. In the next section, we will see a continuous estimation: the density plot.

Approximating density functions

Drawing a **density plot** is a common way to get an idea of the PDF's shape. This plot uses kernel density estimation to create a continuous function that approximates the PDF of our variable distribution. The `StatsPlots` package exports the `density` function, which uses the `KernelDensity` package under the hood. It has two important keyword arguments, `bandwidth` and `trim`, both of which we will discuss in this section. For `histogram`, the `density` function also has a `weights` keyword argument for assigning different weights to data points. The following code creates a density plot using `StatsPlots` with default settings:

```
@df iris density(:PetalLength)
```

The following figure shows the generated density plot:

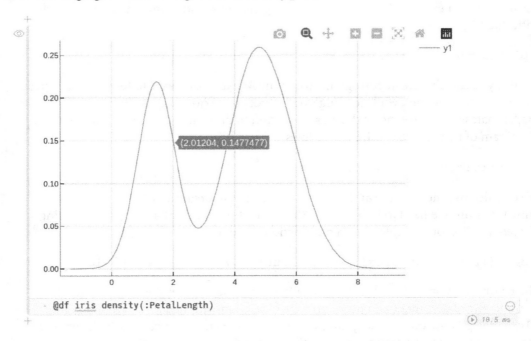

Figure 6.8 – Density plot created in Pluto using StatsPlots and Plotly

As we can see, the density shows us that petal length values under 0 are possible. In our case, this doesn't make much sense, and it is a simple artifact for using a normal distribution as the kernel for our density estimation. The density function offers the `trim` keyword argument to avoid showing a probability higher than 0 outside the data range. In particular, when we set `trim` to `true`, `StatsPlots` trims the curve at the minimum and maximum values of the observed data. For example, let's run the following code:

```
@df iris density(:PetalLength, trim=true)
```

This will create a plot similar to the following:

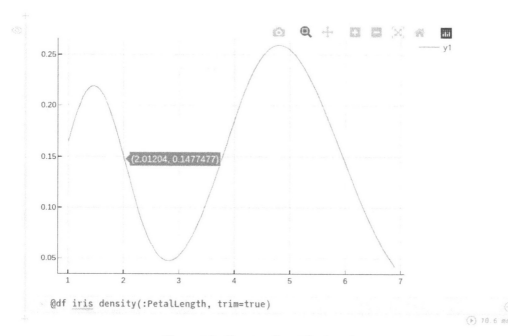

Figure 6.9 – Density plot with trimming

If you compare this plot with *Figure 6.8*, you will notice that while StatsPlots trims the extremes of the curve, the curve height and shape don't change.

We can use the bandwidth keyword argument of the density function to determine the smoothness of the curve. As with histogram bins, the bandwidth value determines the distribution shape we infer from the plot. Large bandwidth values create smooth curves highlighting general trends, while small bandwidths highlight details. By default, StatsPlots uses *Silverman's rule of thumb* to choose the appropriate bandwidth for the plot, depending on the number of data points and their dispersion. This rule could give a wrong bandwidth value when the underlying distribution is far from the normal distribution. Exploring different bandwidth values could help us find the correct one for our visualization. The following code defines two Pluto cells that take advantage of PlutoUI to test different bandwidth values:

```
begin
Pkg.add(name="PlutoUI", version="0.7.23")
    using PlutoUI
    @bind bandwidth Slider(0.1:0.1:10, show_value=true)
end
@df iris density(:PetalLength,
```

```
bandwidth=bandwidth,
label="bandwidth: $bandwidth")
```

The following screenshot shows the interactive interface we created with Pluto:

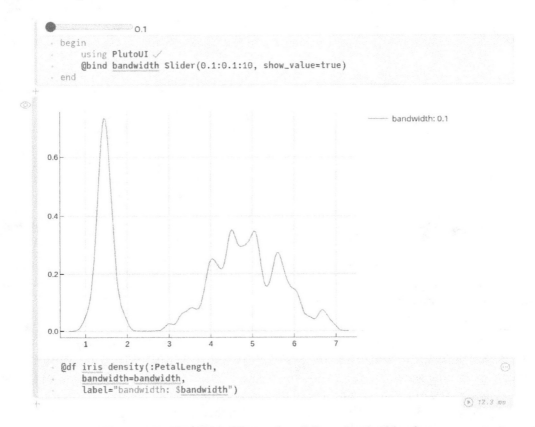

Figure 6.10 – Using PlutoUI to explore different bandwidth values

Now that we have learned how to create density plots with StatsPlots, let's learn how to draw them with Gadfly and Makie. Gadfly offers the density geometry, a combination of the line geometry and the density statistic, to quickly generate density plots. You can create a new Pluto notebook, load Gadfly and the Iris dataset, and run the following code for an example of this:

```
plot(iris, x=:PetalLength, Geom.density)
```

Makie and AlgebraOfGraphics export the density function to create density plots. We have loaded CairoMakie and AlgebraOfGraphics in the same session; therefore, we will use the qualified name of the density functions. The following line of code creates the density plot with Makie:

```
CairoMakie.density(iris.PetalLength)
```

As mentioned previously, we can use the same function with AlgebraOfGraphics using the Density visual:

```
data(iris) * mapping(:PetalLength) * visual(Density) |> draw
```

Finally, we can use the density function defined in the AlgebraOfGraphics package:

```
data(iris) * mapping(:PetalLength) *
      AlgebraOfGraphics.density() |> draw
```

The density function of Gadfly, Makie, and AlgebraOfGraphics also has a bandwidth keyword argument to modify the default bandwidth. As with StatsPlots, all these packages use the KernelDensity package to create the density curve. Therefore, the default behavior is the same for all of them. Some noteworthy differences are that Gadfly doesn't allow trimming and that AlgebraOfGraphics allows you to change the default normal kernel and trims by default.

Now that we have an idea of the data distribution when using approximations of the PDFs, let's visualize distribution descriptors using boxplots.

Introducing boxplots

A **boxplot** or **box-and-whisker plot** is a way to display the location of the **five-number summary**, a non-parametric descriptor of our sample distribution. These five numbers are the values of the following sample percentiles:

- The *0th percentile* is the *minimum value* we have in our data.
- The *25th percentile* or *first quartile* is the value up to which we accumulate 25% of the samples.
- The *50th percentile*, *second quartile*, or **median** is the value that splits our samples into two halves. Therefore, 50% of the samples show a lower value than the median.
- The *75th percentile* or *third quartile* is the point under which we have accumulated 75% of the samples.
- The *100th percentile* is the *maximum* value we observed in our data.

If we have `StatsPlots` loaded, we can run `boxplot(rand(1000000))` to understand how a boxplot displays those values. Since `rand` creates a set of numbers uniformly distributed between 0 and 1, we expect the minimum and maximum values to be close to those extremes. Also, we can expect to have a quarter of the data below 0.25, half below 0.5, and three quarters below 0.75. The following screenshot shows the boxplot that was created:

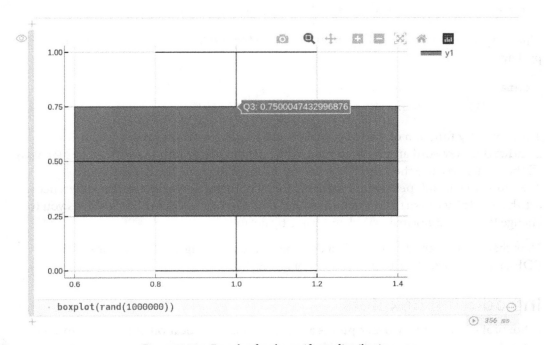

Figure 6.11 – Boxplot for the uniform distribution

As we can see, we can move the mouse pointer close to the line vertices to see the estimated quartiles when using *Plotly* as the backend. Note that the `StatsPlots` package labels the *3rd quartile* as *Q3*.

Let's look at the different parts that compose this boxplot. We can see that the box goes from the *first* to the *third quartile*; therefore, it accumulates 50% of our data samples inside. Inside the box, there is a horizontal line indicating the location of the *median*. In this example, the horizontal lines indicating the first quartile, median, and third quartile are at 0.25, 0.5, and 0.75 on the *y* axis, respectively. There are also two lines, known as the **whiskers**, that go from the box to the minimum and maximum values. In this case, the whiskers extend to the extremes of the data because we do not have outliers to show.

The median is our measure of centrality in a boxplot. Because it is non-parametric, it has the advantage of being more robust than the mean to outliers and skewed distributions. Then, the distance between the first and the third quartile, which is the box's height, represents the **interquartile range** (**IQR**). The IQR is a robust measure of data dispersion and is an alternative to standard deviation.

We have seen that we can create a boxplot using the boxplot function from StatsPlots. In this case, let's plot the distribution of the *sepal widths* using a *notched boxplot*:

```
@df iris boxplot(:SepalWidth, notch=true)
```

In this call to boxplot, we have set the notch keyword argument to true. Therefore, this code creates a boxplot that displays a notch in the box around the median, as shown here:

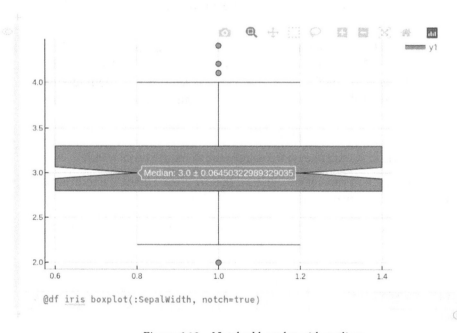

Figure 6.12 – Notched boxplot with outliers

The **notch** indicates an interval around the median. It is proportional to the IQR and inversely proportional to the square root of the number of samples. Therefore, increasing the sample size will decrease the notch's width. These notches are helpful when comparing boxplots; if they do not overlap, then it is likely that the medians are statistically different. As we can see, StatsPlots with the *Plotly* backend will show this interval around the median upon hovering our mouse over it.

There is another thing we can observe in *Figure 6.12*: the presence of **outliers**. `StatsPlots` draws the outliers using points after the whiskers extremes. The limit of the whiskers does not match the extreme data values in the presence of outliers. Instead, the minimum and maximum values are indicated by the first and last outliers, respectively. `StatsPlots` will automatically draw any point that goes beyond 1.5 times the IQR after the box limits as outliers. Note that the whisker will not extend to these thresholds but the most extreme data value inside them. You can use the `whisker_range` keyword argument of the `boxplot` function to change the default `1.5` value. If you set `whisker_range` to `0.0`, then the whisker will extend from the box to the minimum and maximum data values.

There are other helpful keyword arguments you can use with the `boxplot` function, as follows:

- The `outliers` keyword argument takes a Boolean value that determines whether `StatsPlots` will show the outliers. It can help hide the outliers when visualizing very skewed distributions.

- The `whisker_width` argument determines the width of the horizontal line at the end of the whiskers. This keyword argument is set to `:half` by default, indicating that the line should cover half the box. You can also set this argument to `:match` to make the line cover the box entirely. Lastly, you can set it to a positive number that determines the line's total length. Setting it to `0` is a common choice to avoid drawing that line, as shown in *Figure 6.13*.

- You can use the `bar_width` keyword argument to fix the box's width.

- As with histograms, you can set `orientation` to `:horizontal` to draw a horizontal boxplot rather than the default vertical one.

The following code modifies some of the default values for these keyword arguments:

```
@df iris boxplot(:SepalWidth,
    notch=true,
    outliers=false,
    whisker_width=0,
    bar_width=0.2)
```

The preceding code creates the following boxplot:

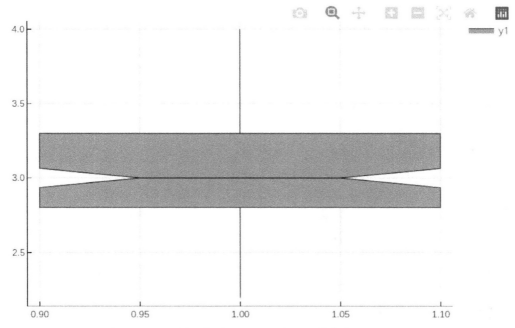

Figure 6.13 – Notched boxplot with non-default arguments

This plot is similar to the one in *Figure 6.12*. However, it hides the outliers, it doesn't show the horizontal line at the extreme of the whiskers, and it has a width of 0.2 units instead of 0.8. You can notice the difference in the box widths by comparing the *x* axis.

Gadfly and Makie also offer a way to draw boxplots. Gadfly draws them using the boxplot geometry. For example, the following code creates a boxplot highlighting the five-number summary for sepal width:

```
plot(iris, y=:SepalWidth, Geom.boxplot)
```

StatsPlots and Gadfly do not need explicit values for the *x* axis. If the *x* values are missing, they will draw the first boxplot at 1.0 on the *x* axis. However, Makie's boxplot implementation needs them explicitly. So, the boxplot function that's exported by the Makie backend needs the first argument to be a vector and indicate the *x* values for each value on the *y* variable. In the following code, we are using broadcasting to apply the one function to each variable element to get a vector of the same length filled with ones:

```
boxplot(one.(iris.SepalWidth), iris.SepalWidth)
```

We can use this `boxplot` function with the `AlgebraOfGraphics` package using the `BoxPlot` visual. In the following example, we have taken another approach to create the vector filled with ones. In this case, we are using the *pair syntax* of `AlgebraOfGraphics` to apply the `one` function to each element in our variable and then renaming the created variable *x*. Note that if you use `AlgebraOfGraphics` 0.6.1 or higher, the `:x` label should be a `String` rather than a `Symbol`; `"x"`, in this example:

```
data(iris) * mapping(
      :SepalWidth .=> one => :x,
      :SepalWidth) *
   visual(BoxPlot) |> draw
```

In this section, we learned how to create boxplots, which are an excellent way to display the five-number summary of our variable. However, boxplots cannot show whether a variable is multimodal or not. The next section will introduce violin plots, which solve that problem.

Visualizing distribution shapes with violin plots

Boxplots can describe a distribution, but they cannot depict the PDF shape. Therefore, we cannot use a boxplot to show that a variable distribution is multimodal. **Violin plots** solve this using a kernel density estimate of the PDF as density plots. We can think of a violin plot as a rotated and reflexed density plot. Let's create one with `StatsPlots` to see an example:

```
@df iris violin(:PetalLength)
```

The following figure shows the violin plot that was created for petal length:

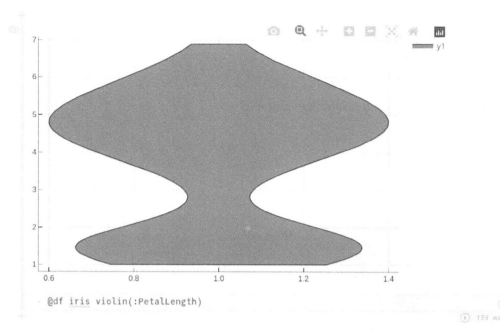

Figure 6.14 – Violin plot

The violin function from the StatsPlots package takes the bandwidth, weights, and trim keyword arguments. They work as the homonymous arguments described for the density function. However, there is a difference because StatsPlots trims violin plots by default. The violin function can also take the following keyword arguments:

- side: StatsPlots sets side to :both by default. Therefore, the density is reflexed at both sides of the vertical line at the violin's center. However, you can set this to :right or :left to have the density plotted only on one side.

- show_mean: StatsPlots will draw a horizontal dashed line at the mean height if you set this Boolean argument to true.

- show_median: This argument will draw a horizontal line at the median position if you set it to true.

- quantiles: You can pass a probability vector to this argument to highlight the respective quantiles. For example, specifying the [0.25, 0.75] vector will make StatsPlots draw horizontal lines at the first and third quartile levels.

For example, in the following code, we have modified some of these keyword arguments:

```
@df iris violin(:PetalLength,
    trim=false,
```

```
    side=:right,
    show_mean=true,
    show_median=true,
    quantiles=[0.25, 0.75])
```

The following figure shows the resulting violin plot:

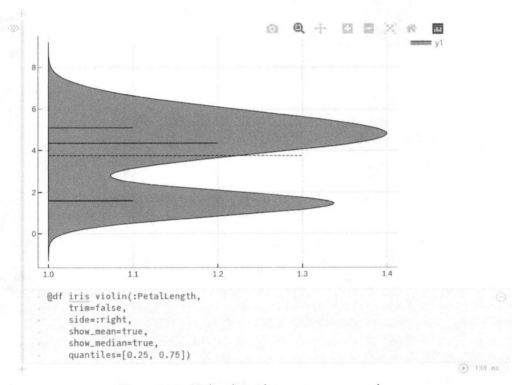

Figure 6.15 – Violin plot with custom argument values

Using the `violin` function, we can use the `show_median` and `quantiles` keyword arguments to show some distribution descriptors on a boxplot. However, if we want them all, a standard solution is to plot a boxplot over a violin plot. To achieve that, we can take advantage of the mutating `StatsPlots` function to define each plot type. For example, while the `violin` function creates a new plot, the `violin!` function will draw a violin plot over an existing plot. In this case, we will use the `violin` function to draw the violin plot in the background and the `boxplot!` function to draw a boxplot over it:

```
begin
    plt = @df iris violin([1], :PetalLength,
```

```
        fillalpha=0.4,
        linewidth=0,
        color=:green,
        legend=false)
@df iris boxplot!(plt, [1], :PetalLength,
        notch=true,
        bar_width=0.1,
        whisker_width=0,
        color=:green)
end
```

Here, we have modified the default value of multiple keyword arguments for aesthetic purposes. The most important is bar_width of boxplot! to avoid overlapping the violin plot entirely. Note that we have to explicitly set the position in the *x* axis – in this case, using the [1] vector – otherwise, StatsPlots will draw the plots in positions 1 and 2. The following figure shows the resulting plot:

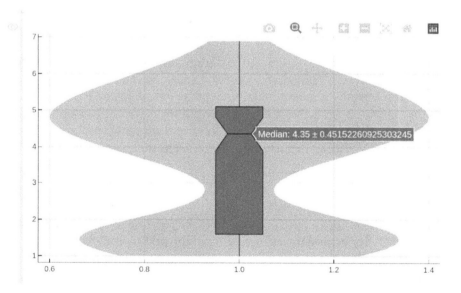

Figure 6.16 – Boxplot drawn over a violin plot

You can also create violin plots using the violin geometry of Gadfly. For example, the following line of code creates a violin plot for petal length using Gadfly:

```
plot(iris, y=:PetalLength, Geom.violin)
```

Makie exports the `violin` function to create violin plots. You can access that function from `AlgebraOfGraphics` using the `Violin` visual. Drawing violin plots with these packages is similar to plotting boxplots; you need to give the values for the *x* axis explicitly. Therefore, to create a violin plot for petal length with Makie, you can use the following code:

```
violin(one.(iris.PetalLength), iris.PetalLength)
```

The equivalent of doing this with `AlgebraOfGraphics` 0.6.0 is as follows:

```
data(iris) * mapping(
    :PetalLength .=> one => :x,
    :PetalLength) *
    visual(Violin) |> draw
```

Now that we have learned how to create violin plots, let's move on to the last statistical plot of this section: the Q-Q plot.

Comparing distributions with Q-Q plots

The **quantile-quantile plot**, **Q-Q plot** for short, is a way to compare two variable distributions. Usually, we want to compare the distribution observed in our sample against a theoretical one. The `StatsPlots` package exports the `qqplot` function to create such plots. This function can take a vector containing our sample data and a `Distribution` object from the `Distributions` package. It also allows us to pass two vectors if we want to compare the quantiles of two samples. Q-Q plots are commonly employed to assess the normality of our data by setting the *normal distribution* as the theoretical one.

Therefore, to make assessing data normality easier, the `StatsPlots` package also exports the `qqnorm` function. Both functions take the `qqline` keyword argument to determine the line to which all the points should align if the distributions are identical. By default, `qqline` is set to `:identity`, drawing the $y=x$ line in the plot. Other options include `:fit`, which performs a linear regression of the points, and `:quantile` or `:R`, which draws a line passing through the first and third quartiles of the distributions. The following line creates a Q-Q plot for sepal length against the normal distribution using default arguments:

```
@df iris qqnorm(:PetalLength)
```

The following figure shows the resulting Q-Q plot:

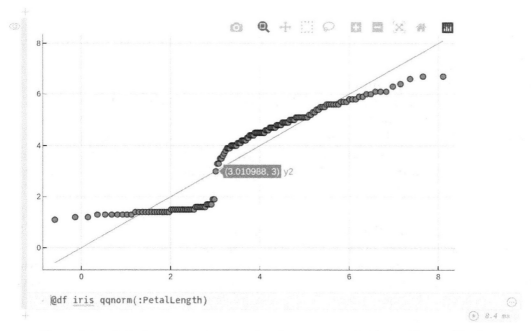

Figure 6.17 – Q-Q plot to assess the normality that was created with StatsPlots and Plotly

Here, the quantiles for the normal distribution are on the *x* axis. Consequently, the quantiles from our sample sepal lengths are on the *y* axis. Here, we can see that our variable distribution is non-normal as the points don't fit the line. Also, the strong s-shape of the curve indicates that our variable distribution is bimodal.

Makie also exports the qqplot and qqnorm functions as the StatsPlots package. Makie functions also take the qqline keyword argument with the same behavior previously described for StatsPlots. When using AlgebraOfGraphics, you need to use the QQPlot or QQNorm object in the visual function.

The Gadfly package offers the qq statistics to draw Q-Q plots; however, it takes more effort to obtain a plot similar to the one created by StatsPlots. The following code will create the Q-Q plot of petal length against the normal distribution using Gadfly:

```
begin
    using Distributions
    plot(layer(iris,
        x=fit(Normal, iris.PetalLength),
        y=:PetalLength,
        Stat.qq),
    layer(identity,
```

```
        extrema(iris.PetalLength)...,
        Geom.line))
end
```

Note that we need to explicitly use the `Distributions` package to access the `Normal` distribution object and the `fit` function. Then, we can fit the normal distribution to our variable to calculate the quantiles for the *x* axis. Also, `Stat.qq` will only draw the points; to get the identity line, we will need to plot it in a separate layer. We can create a `line` geometry for the `identity` function, *y=x*, in the data domain to achieve that. We have used the `extrema` function to get the minimum and maximum data points. As the `extrema` function returns a tuple, we need to use the splat operator, `...,` to pass the tuple's elements as function arguments.

This section has explained the main statistical plots for describing univariate distributions. In particular, we have extensively explored histograms, density plots, boxplots, violin plots, and Q-Q plots using `StatsPlots`. We also learned how to create similar plots with `Gadfly` and the Makie ecosystem. In the next section, we will explore the statistical plots that are available in Julia for visualizing bivariate distributions. We will notice that some of them are two-dimensional versions of the plots we have learned about in this section.

Plotting bivariate distributions and regressions

The easiest way to see how two variables are related is by creating a *scatter plot*, especially with few samples. We can assign one of the variables to the *x* axis and the other to the *y* axis. However, when the number of samples is high, the points overlap, making it hard to know the point density in different plot regions. If the number of points is not too high, adding some *transparency* can alleviate this problem. You can quickly achieve this in `Plots` and `StatsPlots` by setting the `alpha` keyword argument of the `scatter` function to a value that's lower than one (fully opaque) but greater than 0 (fully transparent). Nevertheless, a better way to solve this problem is to create a plot that approximates the *joint probability distribution* of the two variables. The most common ones are the bi-dimensional versions of histograms and density plots.

We can create a **bi-dimensional histogram** using the `histogram2d` function from `Plots` and `StatsPlots`. It is similar to the previously described `histogram` for one variable, but we discretize the two variables defining boxes. Then, we encode the data density in each box using its color. The `histogram2d` function also takes a `bins` keyword argument that works as described previously. However, we can pass a two-element tuple to this argument to choose a different binning strategy for each variable. For example, passing `(5, 10)` to `bins` will tell `StatsPlots` to create approximately five bins for the variable on the *x* axis and 10 for the one on the *y* axis. Bi-dimensional histograms can be easily made with Gadfly using Geom.histogram2d. Within the same package, you can also find the hexbin geometry. With the AlgebraOfGraphics package, you can use the previously described histogram function to provide mappings for two variables, x and y, to get a bi-dimensional histogram.

Bi-dimensional density plots are an excellent way to see the shape of a joint probability density function. There is no shorthand for this plot in Plots or StatsPlots. However, those packages can plot a *kernel density estimation* created with the kde function from the KernelDensity package. The kde function takes a two-element tuple containing the variables. Plots will use a **contour plot** to visualize the estimated density function in two dimensions. Each line of a contour plot represents the shape of the three-dimensional density function cut at a given density value. The line color denotes the density value of the cut. With Makie, you need to plot the result from the kde function using the contour function to get the same plot. Thankfully, Gadfly offers the density2d geometry for creating a contour plot that shows the bi-dimensional density plot easily.

Finally, the AlgebraOfGraphics package also provides an easy way to produce bi-dimensional density plots using the density function. You need to map two variables, x and y, instead of one. The resulting plot indicates the density at each point using a heatmap rather than a contour plot; you can change that by using the Contour visual.

Let's run the following code example to create the three plots we've discussed using StatsPlots:

```
begin
    using RDatasets, StatsPlots
    import KernelDensity
    iris = dataset("datasets", "iris")
    @df iris plot(
        scatter(:PetalLength, :PetalWidth,
            alpha=0.5,
            legend=false, title="Scatter plot"),
        histogram2d(:PetalLength, :PetalWidth,
            bins=(12, 6), color=:grays,
            title="Bi-dimensional histogram"),
        plot(KernelDensity.kde((:PetalLength,
    :PetalWidth)),
            title="Bi-dimensional density plot"),
            xlabel=^(:PetalLength), ylabel=^(:PetalWidth),
```

```
              guidefontsize=8, titlefontsize=10)
   end
```

We needed to import the `KernelDensity` package to create the bi-dimensional density plot from a tuple of variables using the `kde` function. Here, we have set the `alpha` keyword argument of the scatter plot to `0.5` to have points with 50% opacity highlighting their overlap. We have also used the `bins` keyword argument of `histogram2d` so that we have more bins for the x variable.

Another interesting thing to note is that we used `^(:PetalLength)` instead of `:PetalLength` when setting `xlabel`. With `StatsPlots`, inside the `@df` macro, we can use `^()` around an expression containing symbols to escape them and avoid having to replace a variable's name with its values. In this case, we wanted to assign the `:PetalLength` symbol to `xlabel` and not the petal length values. We also used `^()` for `ylabel`.

The following figure shows the plot that was created by the preceding code block:

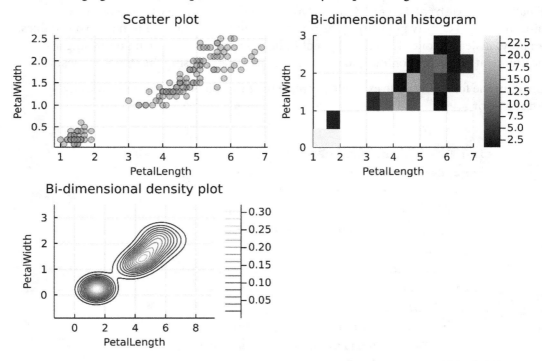

Figure 6.18 – Visualizing the distribution of two variables with StatsPlots

Now that we've learned how to visualize bivariate distributions, let's learn how to draw regression lines to see the association between the variables.

Drawing regression lines

Drawing regression lines is an excellent way to visualize the association between two non-independent variables. `Gadfly` and `AlgebraOfGraphics` offer easy ways to create such plots. There are two kinds of regression lines we can make with these packages. The first is a classical **linear regression** to visualize the linear association between two variables. The second is a **local regression**, usually using a **locally estimated scatterplot smoothing (LOESS)** method. The LOESS method performs polynomial regressions on subsets of the data points. Therefore, they have two important parameters: the **bandwidth** or **smoothing parameter** and the degree of the polynomials. The bandwidth determines the sizes of the data subsets. Therefore, small bandwidth values create regressions that are sensitive to the local variations.

We can create these plots in `Gadfly` using `Stat.smooth`. This function takes a `model` keyword argument to select between a linear model, passing the `:lm` symbol, or a LOESS method, passing `:loess`. Let's explore this function using a Pluto notebook to visualize the association between petal length and width from the Iris dataset:

1. Create a new Pluto notebook.

2. Execute the following code in the first cell:

```
begin
    using RDatasets, Gadfly
    iris = dataset("datasets", "iris")
end
```

The preceding code loads the `Gadfly` package and the Iris dataset.

3. Run the following code block in a new cell to create a *local regression*:

```
plot(iris, x=:PetalLength, y=:PetalWidth,
    layer(Stat.smooth(method=:loess),
        Geom.line))
```

Here, we have indicated that we want to fit a LOESS model by passing `:loess` to the `method` keyword argument of the `smooth` statistics. Also, note that we have created a `layer` that contains the local regression, and we have used the `line` geometry to visualize it.

4. Execute the following code to create a layer with the points:

```
plot(iris, x=:PetalLength, y=:PetalWidth,
    layer(Stat.smooth(method=:loess),
        Geom.line),
    Geom.point)
```

In the preceding code, the point geometry creates a layer under the layer containing the regression.

5. Run the following code block to see the effect of changing the smoothing parameter:

```
plot(iris, x=:PetalLength, y=:PetalWidth,
    layer(Stat.smooth(method=:loess, smoothing=0.2),
        Geom.line))
```

In this case, we have changed the *bandwidth* from its default value of 0.75 to 0.2 using the smoothing keyword argument. This value represents the proportion of data that's been selected for each subset; therefore, it should be a value between 0 and 1. Gadfly uses a second-degree polynomial (quadratic function) for each local fit, so each subset should contain at least three data points. This imposes a lower limit that's higher than 0 for smoothing, depending on the number of points.

6. Execute the following code to create a *linear regression*:

```
plot(iris, x=:PetalLength, y=:PetalWidth,
    layer(Stat.smooth(method=:lm, levels=[0.99]),
        Geom.line, Geom.ribbon))
```

Here, we have passed :lm to the method keyword argument of smooth to create a linear regression. In this case, we also used the line geometry in the regression layer to visualize the regression line. Gadfly allows us to visualize *confidence bands* around a linear regression line. To that end, we can use the levels keyword argument of smooth, which takes a vector of floating-point numbers indicating the desired confidence level; by default, it is 95%. In this case, we have passed a vector containing a value of 0.99 to create a 99% confidence band. To visualize the confidence band, we need to use the ribbon geometry.

The following plot shows the LOESS regression that was created in *Step 3* (on the left) and the linear regression that was created in *Step 6* (on the right):

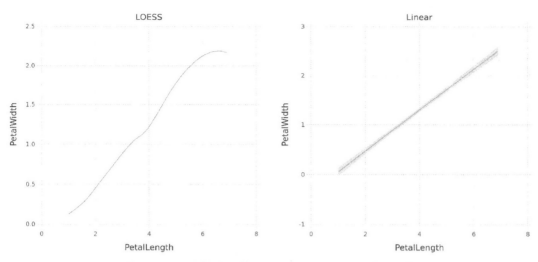

Figure 6.19 – Local and linear regression lines with Gadfly

We can create similar plots in the Makie ecosystem using the linear and smooth functions of the AlgebraOfGraphics package. By default, the linear function creates a *linear regression* line with a 95% confidence band. We can use the interval keyword argument to define whether we want to draw a *confidence band* or a *prediction band* that determines bounds for the predicted values. We need to pass the :confidence symbol to linear for the former and :prediction for the latter. The AlgebraOfGraphics package implementation does not allow us to choose a different confidence level but does allow us to create *weighted linear regressions*. To get a weighted linear regression, you need to define weights in the mapping function.

The smooth function of the AlgebraOfGraphics package creates a *local regression* line using the LOESS method. This function takes two keyword arguments – span to set the smoothing parameter and degree to specify the polynomial degree. The span keyword argument can take values between 0 and 1 and is set to 0.75 by default. The AlgebraOfGraphics package also uses a degree of 2 by default.

In this section, we learned how to create linear and local regression lines using Gadfly and AlgebraOfGraphics. In the next section, we will learn how to draw marginal plots to visualize bivariate and univariate distributions conjointly.

Creating marginal plots

In the previous sections, we learned how to visualize univariate and bivariate distributions using StatsPlots. In this section, we will learn how to create a plot that simultaneously displays the joint and *marginal distributions* for two variables. Usually, we can achieve this by showing three plots in the same figure. The central plot shows the bivariate distribution, for example, using a bi-dimensional histogram. Then, two small plots at the top and right-hand sides of the central plot show the univariate distributions for the *x* and *y* variables, respectively. We can take advantage of the layout capabilities of the Plots package to create such visualizations. We briefly introduced those capabilities in the *Simple layouts* section of *Chapter 1, An Introduction to Julia for Data Visualization and Analysis*, and we will expand on that in *Chapter 11, Defining Plot Layouts to Create Figure Panels*. However, StatsPlots also offers a series of recipes for marginal plots that make it easier to create this kind of visualization.

The StatsPlots package offers three functions for visualizing the joint and marginal distributions of two variables. The first is marginalhist, which creates a plot similar to the following:

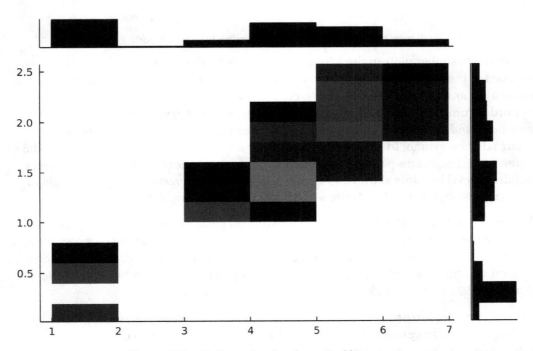

Figure 6.20 – Bi-dimensional and marginal histograms

As we can see, the central plot is a bi-dimensional histogram that shows the joint probability distribution of the *x* and *y* variables. Then, the *marginal histograms* highlight the marginal distributions of *x* and *y*. We can use the bins keyword argument in the same way we have described for the histogram2d function to determine the central plot's binning. Then, the marginal histogram's binning will match the binning of the bi-dimensional histogram in the corresponding dimension. The marginalhist function also takes a `density` keyword argument. `StatsPlots` will use density plots to show the marginal distributions instead of histograms if we set `density` to `true`.

The second function that `StatsPlots` offers to create these plots with marginal distributions is `marginalscatter`, which draws a scatter plot in the center and two marginal scatter plots. It also takes a `density` keyword argument to plot marginal density plots instead.

Finally, the third function is `marginalkde`. It draws a bi-dimensional density plot in the center and two density plots at the top and the right to show the marginal distributions. The `marginalkde` function takes a `levels` keyword argument to determine the number of contour lines in the main plot. We can also use the `clip` keyword argument to determine the minimum and maximum values for the x and y axes. The `clip` keyword argument takes a tuple of tuples, and `StatsPlots` sets it to `((-3.0, 3.0), (-3.0, 3.0))` by default. Those numbers are multiplied by the 16th and 84th percentiles of the x and y variables to define their limits.

The three functions described here are suitable for displaying the distribution of two variables. If we deal with more variables, we can take advantage of the `corrplot` and `cornerplot` functions of `StatsPlots`. The `corrplot` function creates a correlation plot from a series of variables. The resulting figure has a matrix layout where the lower triangle shows scatter plots for each pair of variables. `StatsPlots` colors each scatter plot according to the *Pearson correlation coefficient* and draws a linear regression line on top of them. Then, the diagonal shows a histogram for each variable. Finally, the upper triangle shows bi-dimensional histograms for the variable pairs. For example, we can execute the following code to create a correlation plot with Julia:

```
begin
    using RDatasets, StatsPlots
    iris = dataset("datasets", "iris")
    @df iris corrplot(
    [:SepalLength :SepalWidth :PetalLength :PetalWidth],
    fillcolor=cgrad(), guidefontsize=9, tickfontsize=6)
end
```

The following figure shows the plot that was created:

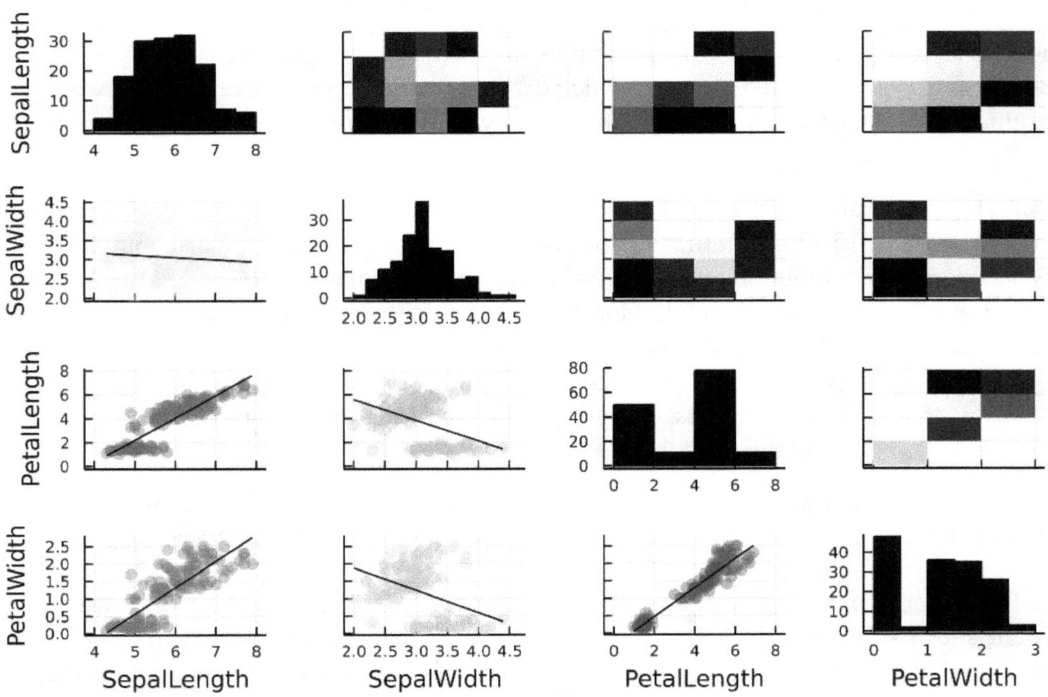

Figure 6.21 – Correlation plot created with StatsPlots

The `cornerplot` function produces a similar plot to the one shown in *Figure 6.21*, but the upper and lower triangles are symmetric, showing the same scatter plots. The diagonal also contains scatter plots, and the plot shows marginal histograms for each row and column. You can set the `compact` keyword argument to `true` to plot only the lower triangle and the marginal histograms.

With that, we've learned how to visualize the univariate and bivariate distributions of multiple variables using `StatsPlots`. Now, let's learn how to display clustering results.

Visualizing clustering results

The `Clustering` package offers multiple clustering algorithms for Julia. These algorithms aim to create groups where the elements in a group are more similar than those between groups. In particular, it provides the `hclust` function for performing a *hierarchical clustering* from a distance matrix. It creates a *dendrogram*, where the most similar data points, known as the leaves, are closer. This means you need to travel fewer and shorter branches from one leaf to another that belong to the same cluster than from visiting leaves outside it. The function returns an object of the `Hclust` type. The `StatsPlots` package exports a `Plots` recipe to draw the *dendrogram* when you call the `plot` function with a `Hclust` object. Let's create a dendrogram that clusters the variables in the Iris dataset:

1. Create a new Pluto notebook.

2. Execute the following code in the first cell to load the necessary libraries and the Iris dataset:

   ```
   begin
       using Clustering, Distances
       using RDatasets, StatsPlots
       iris = dataset("datasets", "iris")
   end
   ```

   We will use the `Distances` package to create the required distance matrix.

3. Run the following code:

   ```
   distances = pairwise(CorrDist(), Matrix(iris[!, 1:4]))
   ```

   Here, we used the `pairwise` function to measure the distances between matrix columns. In particular, we estimated the correlation distance between the variables in the Iris dataset, as indicated by the `CorrDist()` object from the `Distances` package. Because the `pairwise` function needs to take a matrix, we used `Matrix` to convert the first four columns of our DataFrame into a matrix.

4. Execute the following code to perform the clustering:

   ```
   clustering = hclust(distances, branchorder=:optimal)
   ```

   Here, we created a hierarchical clustering from the distance matrix using the `hclust` function from the `Clustering` package. We set the `branchorder` keyword argument to `:optimal` to order the tree branches to reduce the distance between neighboring leaves.

5. Run `plot(clustering)` to draw the dendrogram for the clustering results. The leaves' names in the *x* axis are simply the column numbers in the distance matrix; that is, the column index in our original DataFrame. Let's label the leaves of the dendrogram.

6. Run `variables = names(iris)[1:4]` to create a vector with the variable names in the original table.

7. Execute the following code block:

```
plot(clustering,
     xticks=(1:4, variables[clustering.order]))
```

Here, we set `xticks` to label the leaves. The `xticks` keyword argument needs a two-element tuple. The first element indicates the position in which we want to put the labels, while the second indicates the labels to use. We used the `order` attribute of the `Hclust` object, which contains the indices of the distance matrix columns in the order they appear in the leaves. Therefore, we can use `order` to index the vector with the variable names to get them in the correct order. The preceding code produces the following plot:

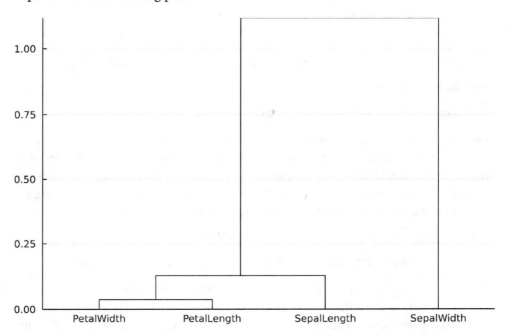

Figure 6.22 – Dendrogram created with Clustering and StatsPlots

8. Run the following code:

```
cutree(clustering, k=2)
```

Here, we used the `cutree` function to assign each variable from the Iris dataset to a cluster. This function takes an `Hclust` object and the k or h keyword argument. k defines the expected number of clusters, while h defines the height at which we must cut the dendrogram to define the clusters. In this case, we obtained the [1,

`2, 1, 1]` vector, meaning that the second variable, following the order in the distance matrix (*SepalWidth*), belongs to cluster 2. In contrast, all the other variables belong to cluster 1. We have obtained two clusters because we have set k to 2.

We need to use the `assignments` function to get a similar vector while assigning variables or observations to cluster numbers when we use other clustering methods from the `Clustering` package. The vector that's returned from the `cutree` or `assignments` function can help us visualize the clustering result by grouping or coloring each cluster. We will discuss how to perform such operations with `StatsPlots` in the next section.

Comparing between groups

Groups and clusters are usually defined using categorical or ordinal variables. For example, `Species` is a categorical variable grouping observation according to the species name in the Iris dataset. The `assignments` and `cutree` functions of `Clustering` return a vector with integers for encoding a categorical or ordinal variable. We consider that variable to be ordinal if the order of clusters has a meaning, for example, if they are related to the number of elements in the cluster. However, we will always visualize the assignation vector as a categorical variable when discriminating clusters.

We can use the *position*, *color hue*, and *shape* aesthetics to visualize and discriminate groups when using packages based on the Grammar of Graphics. *Facets* are also an excellent way to compare the same plot between groups. In *Chapter 5*, *Introducing the Grammar of Graphics*, we learned how to map categorical variables to those aesthetics and create facet plots using `Gadfly` or the `AlgebraOfGraphics` package. In this section, we will visualize categorical variables using `StatsPlots`. Let's open a new Pluto notebook to explore this using the `Species` variable of the Iris dataset:

1. Execute the following code block in the first cell:

    ```
    begin
        using RDatasets, StatsPlots
        iris = dataset("datasets", "iris")
    end
    ```

2. Run the following code in a new cell:

    ```
    @df iris boxplot(:Species, :PetalLength)
    ```

    Here, we used the `:Species` categorical variable to define the position of each group on the *x* axis. Having each boxplot side by side makes it easy to compare them. Using the first positional argument to determine the group and compare distributions side by side works well for the `boxplot`, `violin`, and `dotplot` functions.

3. Execute the following code block in a new cell:

```
@df iris density(:PetalLength, group=:Species)
```

Here, we created three density plots, one per species. The `group` keyword argument splits the input data into different series matching the categories. Then, `Plots` assign a different color to each series to identify each group.

4. Run the following code in a new cell:

```
@df iris groupedhist(:SepalLength, group=:Species)
```

Using the `group` keyword argument with the `histogram` function will create overlapped histograms with different bins that can hide important trends. To solve that problem, `StatsPlots` exports the `grouphist` function, which plays well with grouping. By default, `grouphist` places histogram bars side by side. This is equivalent to setting the `bar_position` keyword argument to `:dodge`.

5. Execute the following code in a new cell:

```
@df iris groupedhist(:SepalLength, group=:Species,
    bar_position=:stack)
```

Here, we have used the `grouphist` function again, but we have set the `bar_position` keyword argument to `:stack` to create a stacked histogram. As you can see, by comparing the plots of the last three cells, density plots are better than histograms when comparing the shape of overlapped distributions. You can see these three plots in the following figure – they are in different colors:

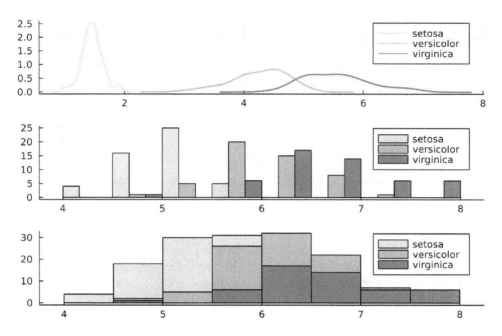

Figure 6.23 – Comparing three distributions using a density plot, a histogram, and a stacked histogram

6. In a new cell, execute the following code:

```
@df iris scatter(:PetalLength, :PetalWidth,
    group=:Species)
```

Here, we used the `group` argument of `scatter` to create three series of points, each with a different color.

7. Execute the following code in a new cell:

```
@df iris scatter(:PetalLength, :PetalWidth,
    group=:Species,
    markershape = [:x :circle :square])
```

Here, we passed the 1x3 `[:x :circle :square]` matrix the `markershape` keyword argument of `scatter` to assign a different shape to each series created by `group`.

With that, we've looked at the essential elements for comparing groups or clusters using `StatsPlots`. While we used the `Species` variable of the Iris dataset for the examples, you can use another vector or collection to assign a group to each observation. For example, you can group according to the output of the `assignments` or `cutree` functions of the `Clustering` package. What's more, the `group` keyword argument can also take a tuple of variables to break the data accordingly by combining those variables.

Summary

In this chapter, we learned how to use `StatsPlots`, the library for drawing statistical plots in the `Plots` ecosystem. First, we learned how to visualize univariate and bivariate distributions using `StatsPlots`, `Gadfly`, and `AlgebraOfGraphics`. Then, we explored the basic statistical plots that we can combine to display joint and marginal distributions. Finally, we learned how `StatsPlots` can compare distributions across groups and visualize clustering results. What you've learned in this chapter, combined with the knowledge you acquired in *Chapter 1, An Introduction to Julia for Data Visualization and Analysis*, and *Chapter 5, Introducing the Grammar of Graphics*, will enable you to make the most of Julia for visualizing and exploring tabular data. In the next chapter, we will learn about the visualization of another data layout: graphs.

Further reading

Take a look at the following resources if you want to learn more about how to perform statistical visualization with `Gadfly` and `AlgebraOfGraphics`:

- Gadfly: `http://gadflyjl.org/stable/`
- AlgebraOfGraphics: `http://juliaplots.org/AlgebraOfGraphics.jl/stable/`

7
Visualizing Graphs

We commonly use graphs to represent a set of entities and their relationships. Therefore, they are pretty valuable for many areas of knowledge, such as social and computer sciences.

This chapter will introduce us to the main Julia packages for analyzing and visualizing graphs. First, we will learn how to create graph objects using the `Graphs` and `MetaGraphs` packages. Then, we will learn about the graph visualization packages available in the Julia ecosystem. We will focus on the `GraphRecipes` package from the Plots ecosystem. We will also briefly introduce the `GraphMakie` and `GraphPlot` packages. Next, we will expand on the different graph layouts available for those libraries. Finally, we will showcase some standard analyses on graphs and ways to visualize them. By the end of this chapter, you will know how to start analyzing and plotting graphs with Julia.

In this chapter, we're going to cover the following main topics:

- Working with graphs in Julia
- Graph visualization packages
- Graph layouts
- Analyzing graphs

Technical requirements

For this chapter, you will need Julia, Pluto, and a web browser with an internet connection. In the `Chapter07` folder of this book's GitHub repository, you will find the Pluto notebooks that contain this chapter's code examples and their HTML versions, along with their outputs: `https://github.com/PacktPublishing/Interactive-Visualization-and-Plotting-with-Julia`.

Working with graphs in Julia

Graphs are mathematical structures that represent the relationship between objects. A graph is a tuple of two sets; the first contains the **vertices** representing the objects, while the second contains the **edges** describing the relationship between vertex pairs. We can also refer to graphs as **networks**, especially when they depict real-world systems. In that case, we can call vertices **nodes** and edges **links**. As networks represent real objects, it is common to have attributes; for example, labels for their nodes and links. We usually plot graphs using dots to represent the vertices and lines to represent the edges; therefore, it is also common to refer to vertices as **points** and edges as **lines**.

We can define different graph types according to the number and direction of the represented relationship between vertices. A **simple graph** has no more than one edge between vertex pairs. A graph with more than one edge between vertices – for example, one that has different edges representing different relationships – is called a **multigraph**. We can also classify graphs as **directed** or **undirected**, depending on whether their edges have an orientation. A *directed graph* has directed edges, usually represented as arrows. If a directed edge goes from vertex x, the **source** vertex, to vertex y, the **destination** vertex, we say that x is the **tail** of the edge and y is its **head**. For example, we can represent a citation network using a simple directed graph, where edges go from the citing article to the cited one. However, a collaboration or co-authorship network will be undirected, as there is no direction in that relationship.

The `Graphs` package is the central package of the *JuliaGraphs* organization. It defines the `AbstractGraph` type and the basic interface for Julia's graphs that other packages from the ecosystem can extend. The `Graphs` package also exports the `SimpleGraph` type for representing *simple undirected graphs* and `SimpleDiGraph` for *simple directed graphs*. Those types do not store attributes for vertices and edges; if you need that, you can use the `SimpleWeightedGraphs` or `MetaGraphs` packages from the *JuliaGraphs* ecosystem. The first package exports the `SimpleWeightedGraph` and `SimpleWeightedDiGraph` types to represent simple undirected and directed graphs, respectively, and store their edge weights. These **weights** are simple numerical values whose meaning depends on the described system; for example, a collaboration network could have weights indicating the number of shared publications. However, adding or removing edges and vertices to graphs defined using the `SimpleWeightedGraphs` package can be slow. Those operations that modify the graph are more performant on objects from the `MetaGraphs` package. The `MetaGraphs` package exports the `MetaGraph` and `MetaDiGraph` types to represent simple undirected and directed graphs, respectively, with edge weights and arbitrary metadata for vertices and edges. Therefore, `MetaGraphs` is also a better choice than `SimpleWeightedGraphs` if we want to store more complex data for vertices and edges.

With that, we have briefly introduced the different kinds of graphs we can have. We have also mentioned the different types in the Julia ecosystem that we can use to represent them. Now, let's see them in action to learn about the fundamental operations we can perform on them. We will focus on the `Graphs` package, and we will use the `GraphRecipes` package from the Plots ecosystem to explore them visually. Let's create a new Pluto notebook and perform the following steps:

1. In the first cell of the notebook, execute the following code:

    ```
    using Graphs, GraphRecipes, Plots
    ```

    Here, we have loaded the `Graphs` package for creating and working with graphs and the `GraphRecipes` and `Plots` packages to plot them.

2. In a new cell, execute the following code:

    ```
    graph = SimpleGraph(3)
    ```

    Here, we have used the `SimpleGraph` constructor to create a *simple undirected graph* with three vertices and no edges. Note that the number that's used to indicate the desired vertices should be an integer. We can also give a second number to the constructor to let it create that number of random edges. Let's check the number of vertices and edges in the graph we have just created.

3. In a new cell, execute `nv(graph)` to see the *number of vertices* in our graph.

4. Execute `ne(graph)` in a new cell to see the *number of edges* in our graph.

5. In a new cell, execute the following code:

    ```
    typeof(graph)
    ```

In a 64-bit machine, the type of our graph will be SimpleGraph{Int64}, while in a 32-bit machine, it will be SimpleGraph{Int32}. As you can see, the SimpleGraph type is parametric, and the type parameter indicates the type used to store the vertices. By default, it will be Int32 or Int64, depending on your machine. If you create the graph using a number to indicate the number of vertices, as we did in *step 2*, the type parameter will match its type. Note that integer literals will also be of the Int32 or Int64 type, depending on your machine, by default.

6. Execute the following in a new cell:

```
collect(vertices(graph))
```

The vertices function returns an iterator over the graph's vertices at the point. Then, we create a vector containing the vertex values using the collect function. As we can see, the types in the Graphs package represent the vertices using *integers* from 1 to the number of vertices. At the moment, we do not have edges between these three vertices; let's add some.

7. Execute the following in a new cell:

```
add_edge!(graph, 1, 2)
```

Here, we have used the add_edge! function to *add an edge* from vertex number 1 to vertex number 2 in our graph. As you can see, the function has returned true. This means that we have correctly added the edge to our graph. If the edge exists in the graph, the function will return false as SimpleGraph instances don't support multiple edges between vertices. If you want, you can test this by re-running this cell again. The add_edge! function will also return false if the indicated vertex numbers don't exist in our graph.

8. Execute add_edge!(graph, 1, 3) in a new cell to add an edge from vertex 1 to vertex 3.

9. In a new cell, execute the following code:

```
graphplot(graph)
```

Here, we used the graphplot function from GraphRecipes to visualize our simple undirected graph. As we can see, our graph now has three vertices and two edges. The following figure shows the generated plot; note that the exact vertex positions will change between runs:

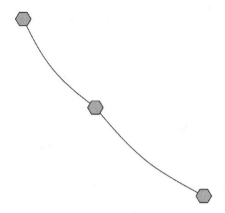

Figure 7.1 – Simple undirected graph plotted with GraphRecipes

10. Execute the following code in a new cell:

```
add_vertex!(graph)
```

Here, we have used the `add_vertex!` function to *add a new vertex* to our graph. The `add_vertex!` function returns `true` if the addition has been successful. It will return `false` otherwise, for example, if we add more vertices than our type can store.

11. Execute the following code in a new cell:

```
add_vertices!(graph, 2)
```

Here, we have used the `add_vertices!` function to add a specific number of vertices to one operation. The function takes the graph to modify and the number of vertices to add, and it will return the number of successfully added vertices.

12. Run `nv(graph)` in a new cell. We will see that our graph now has six vertices; three added upon it being created, one added with `add_vertex!`, and two added using the `add_vertices!` function.

13. In a new cell, execute the following line of code:

```
rem_vertex!(graph, 4)
```

Here, we have used the `rem_vertex!` function to *remove a vertex* – in this case, vertex 4, which we added in *step 10*. Note that deleting a vertex means swapping the vertex to remove with the last vertex and then removing the last. Therefore, the operation changes the number associated with those vertices, so we should be cautious when storing vertices' metadata in external data structures using the vertex numbers. You can avoid that problem by using the `MetaGraphs` package's types to store the metadata in the graph directly. The `rem_vertex!` function returns `true` if the deletion was successful.

14. Execute `collect(vertices(graph))` in a new cell. We only have five vertices now since we deleted one vertex in the previous step. The previous step deleted vertex 4 by swapping vertices 4 and 6 and then deleting the last one. Therefore, in this example, the previous vertex, 6, is now vertex 4, and vertex 5 is the last now that previous vertex 4 has gone. As you can see, keeping track of vertices using vertex numbers can be tricky if you perform mutation operations on the graph.

15. Execute `rem_edge!(graph, 1, 3)` to *remove the edge* linking vertices 1 and 3. The `rem_edge!` function returns `true` when the edge has been successfully deleted.

With that, we've learned how to add and remove vertices and edges using a simple undirected graph. Now, let's create a new Pluto notebook to explore directed and undirected graphs using the `Graphs` package:

1. In the first cell of the notebook, execute `using Graphs, GraphRecipes, Plots` to load the necessary packages.

2. In a new cell, execute the following code:

```
directed = SimpleDiGraph(3, 4, seed=1)
```

Here, we have randomly created a *simple directed graph* with three vertices and four edges using the `SimpleDiGraph` constructor. We have decided to start with a determined edge number, though we can also create a directed graph with no edges by passing only the number of vertices to the constructor. We have used the `seed` keyword argument to set up the random number generator to ensure we see the same graph.

3. Execute the following code in a new cell to visualize our new graph:

```
plot_directed = graphplot(
    directed,
    names=vertices(directed),
    nodecolor=:white)
```

Here, we have used the `graphplot` function, as we did previously, but this time, we have changed the vertices colors and added labels to them. We have assigned labels to the vertices using the `names` keyword argument. This argument takes a vector containing the labels we want to give to each vertex; in this case, we used the `vertices` function to label them using their integer number. Then, we changed the default blue color of the *nodes* or *vertices* to white using the `nodecolor` keyword argument to make the vertex number easier to read. As we plot a directed graph, `GraphRecipes` represents the edges using arrows, which indicate the edges direction (*Figure 7.2*).

4. Execute the following code in a new cell:

```
undirected = SimpleGraph(directed)
```

Here, we have created a *simple undirected graph* from our simple directed graph using the `SimpleGraph` constructor.

5. Execute the following code in a new cell:

```
plot_undirected = graphplot(
    undirected,
    names=vertices(undirected),
    nodecolor=:white)
```

Like that, we have plotted the undirected version of our graph.

6. In a new cell, run the following code:

```
plot(plot_directed, plot_undirected)
```

Here, we have plotted the directed and undirected versions of the same graph side by side. *Figure 7.2* shows the created plot. Let's see vertex 3; it has two edges, one coming from vertex 1 and another from 2, and one going out toward vertex 1 in the *simple directed graph*. However, we only have two undirected edges in the *simple undirected graph* associated with vertex 3. Let's find that information using the functions that the `Graphs` package offers:

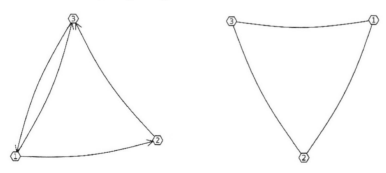

Figure 7.2 – Simple directed and undirected graphs

7. Execute the following code in a new cell:

```
neighbors(undirected, 3)
```

The `neighbors` function lists neighboring vertices from a given vertex. In particular, it returns a list of the vertices that can be reached from the indicated vertex. In this case, we can see that vertex 3 is linked to vertices 1 and 2.

8. Execute the following line of code in a new cell:

```
neighbors(directed, 3)
```

You will see that this returns a list containing only vertex 1. That's because there is only a single edge going out from vertex 3. Therefore, this function produces the same output as the `outneighbors` function for directed graphs, which return the vertices linked by outgoing edges. If you want to get a list of all the vertices connected to a given vertex, regardless of the direction in a directed graph, you should call the `all_neighbors` function.

9. Execute `inneighbors(directed, 3)` in a new cell. The `inneighbors` function returns a list containing all the vertices that have vertices coming to a given vertex. In this case, vertex 3 has edges coming from vertices 1 and 2.

10. In a new cell, execute the following code line:

```
has_edge(undirected, 3, 2)
```

The `has_edge` function takes a graph and two vertices to determine if an edge exists in the graph; the first vertex indicates the source of the edge and the second its destination. Therefore, this function considers the direction of the edge for the directed graph. Let's see an example.

11. In a new cell, execute the following code:

```
has_edge(directed, 3, 2)
```

Here, we can see that the function returned `false` since the only edge linking the two vertices in the directed graph is vertex 2 as the source and 3 as the destination.

In this section, we saw the main operations that we can perform on graphs: adding and removing vertices and edges, testing for the presence of a given edge, and listing neighboring vertices from a vertex. The `Graphs` package also exports functions for creating random graphs and algorithms to traverse graphs and find the shortest paths. Describing the whole set of functionality offered by the `Graphs` package goes beyond the scope of this book. However, we will see a little more in the following sections.

Before you start plotting, let's learn how to load and work with networks using the MetaGraphs package.

Working with MetaGraph

When working with networks, we will find the MetaGraphs package helpful. The package exports the MetaGraph and MetaDiGraph types. Those types store graph, edge, and vertex attributes or metadata using dictionaries. These dictionaries contain keys of the Symbol type, but their values could be of any type. Therefore, we use symbols to name the properties. There is one exception: the MetaGraphs types store the *edge weights* differently, so you will not see the default weights when listing all the edge properties. However, we can set the :weight property of an edge to change its value. In that case, the :weight property will appear in the dictionary of properties for that edge. You can obtain the weights matrix and all the edge weights using the weights function.

Using the type constructors, you can create an instance of the MetaGraph or MetaDiGraph types from a SimpleGraph or SimpleDiGraph object. Note that the type constructors from the MetaGraphs package differ from the ones of the Graphs package. Specifically, if we give two numbers to a MetaGraph constructor, the second number will indicate the default edge weights rather than the edge number. As with the Graphs constructors, the first number we pass in the arguments will determine the number of vertices.

Using the set_prop! function, we can easily *set the properties* of graphs, edges, and vertices. There is also a set_props! function, which takes a dictionary to set multiple properties in a single call. Then, you can *access the properties* using the get_prop function. This function will give you the value of a particular property; if you want to get the dictionary that contains all the properties, you need to use the props function. Other functions that can be useful are the has_prop function, to test whether a particular property exists for the selected element, and the rem_prop! function, to remove a property.

All these functions have a similar signature; the first argument should be a MetaGraph or MetaDiGraph. Then, the following arguments will determine whether we are working with the properties from the graph, vertex, or edge. Namely, if we do not give any extra argument, we will work with the properties of the graph. If we only provide one number, that number will indicate the vertex to be selected. Then, we can pass two numbers to choose an edge, where the first number will denote the source vertex and the second the destination vertex. After those numbers, some functions can take an extra argument to indicate the property we want to use. For that, we will use a Symbol object. The set_prop! function can also take an additional argument containing the value to assign to the given property. And, as we mentioned previously, we can pass a dictionary that goes from property names to values to `set_props!` instead of a single `Symbol` and value to assign multiple properties.

Let's look at a quick example of some of these functions in action in a new Pluto notebook:

1. Execute the following code to load the necessary libraries in the first cell:

    ```
    begin
        using Graphs, MetaGraphs
        using Plots, GraphRecipes
    end
    ```

    Here, we have loaded the `Graphs` and `MetaGraphs` packages to work with `MetaGraphs`. Then, we loaded `Plots` and `GraphRecipes` to plot our graph.

2. In a new cell, execute the following code:

    ```
    graph = path_graph(3)
    ```

    Here, we have used the `path_graph` function from the `Graphs` package to create a *path graph*, also known as a *linear graph*, with three vertices. The `Graphs` package exports many other functions to create random and classic graphs.

3. Execute `graphplot(graph)` to visualize the created graph; as you can see, it is similar to the one in *Figure 7.1* as it also has three vertices and two edges.

4. In a new cell, execute the following line of code:

    ```
    network = MetaGraph(graph)
    ```

    Here, we created a `MetaGraph` from our `SimpleGraph` using the `MetaGraph` constructor. As you will note in the cell's output, the network stores the edge weights using the `:weight` property with a weight of `1.0` for each edge by default.

5. Execute the following code in a new cell:

    ```
    edge_weights = weights(network)
    ```

    Here, we have used the `weights` function to get the `MetaWeights` object containing the *edge weights*.

6. In a new cell, run the following line of code:

    ```
    set_prop!(network, 1, 2, :weight, 0.5)
    ```

    Here, we have set the `:weight` property of the edge going from vertex 1 to vertex 2 of our network to `0.5` using the `set_prop!` function.

7. In a new cell, run the following code:

```
collect(edge_weights)
```

Here, we have collected all the edges in a matrix using the `collect` function. As the graph is undirected, the resulting matrix is symmetrical. As we can see, all edges weigh 1.0 except for the edge linking vertex 1 and 2, which weighs 0.5. The matrix rows indicate the source vertices for directed graphs, while the columns indicate the destinations.

8. Execute the following in a new cell:

```
edge_weights[1, 2]
```

Earlier, we indexed the `MetaWeights` object as a matrix, where the first dimension is the source vertex of the edge and the second its destination. In this case, we accessed the weight of the edge linking vertices 1 and 2.

9. In a new cell, run the following code:

```
set_prop!(network, :name, "Example network")
```

Here, we used the `set_prop!` function to set the `:name` property of the *graph*. We do not give any numerical argument between the graph and the property key to indicate that we want to set a property for the entire graph. The last argument is the value we want to assign to the property; in this case, a `String`.

10. In a new cell, run the following code:

```
set_prop!(network, 1, :name, "First node")
```

Now, we have given a vertex number after the graph to set the property of the indicated *vertex*. In this case, we set the `:name` property of vertex 1. The `set_prop!` function returns `true` because the vertex exists.

11. In a new cell, run the following code:

```
set_prop!(network, 1, 2, :name, "First link")
```

Here, we have given two vertex numbers to set the property of an *edge*. In this case, we have set the `:name` property of the edge that contains vertex 1 as a source and vertex 2 as a destination. The `set_prop!` function returns `true` because the edge exists.

12. In a new cell, execute the following line of code:

```
get_prop(network, :name)
```

Here, we have used the `get_prop` function to get the value of the `:name` property of the graph.

13. In a new cell, execute the following code:

```
get_prop(network, 1, :name)
```

Here, we have accessed the `:name` property of vertex `1`.

14. In a new cell, execute the following line of code:

```
get_prop(network, 1, 2, :name)
```

Here, we get the value of the `:name` property of the edge linking vertex `1` and `2`.

15. In a new cell, run the following code:

```
props(network, 1, 2)
```

Here, we used the `props` function to get the dictionary with the properties of the edge linking vertices `1` and `2`.

Now, we know the basics of setting and getting property values for graphs, edges, and vertices using `MetaGraphs`. This will be useful when visualizing networks. Now, let's learn how to save and load `Graphs` and `MetaGraphs`.

Saving and loading graphs

The `Graphs` package offers three functions for saving and loading graphs: `savegraph`, `loadgraph`, and `loadgraphs`. The `savegraph` function saves one or multiple graphs into a file on disk. The first argument for this function is the file's name. The second argument will depend on the number of graphs we want to store in the file. If we're going to save multiple graphs in a single file, we need to pass a dictionary containing graph names on its keys and graph objects in its values. However, if we're saving a single graph, we must provide the graph object. Optionally, we can use a third positional argument to indicate the desired graph name. Note that graph names should be of the `String` type and that not all formats support storing multiple graphs.

By default, `savegraph` writes the file using the *LG format* from the `Graphs` package. This format can hold one or more graphs in a file. You can use the last positional argument of the `savegraph` function to choose another file format. If you want, you can use the `GraphIO` package to save and load graphs in other formats, such as *GraphML* or *Pajek NET*.

Once you have saved your graph, you can load them using the `loadgraph` function. The first positional argument of `loadgraph` indicates the file to read. The second argument is optional and determines the name of the file we want to read. The last argument is also optional and denotes the file format. By default, the function reads graphs in the *LG format*. If the file format supports storing multiple graphs, you can use the `loadgraphs` functions to load them all.

Let's save and load a graph in *GraphML* format in a `Pluto` notebook to get a taste of this process:

1. Execute the following code in the first cell of the notebook:

    ```
    using Graphs, EzXML, GraphIO
    ```

    `Graphs` can only save and load graphs in *LG format*. To use the *GraphML* format, we must load the `EzXML` package and then the `GraphIO` package. The latter needs the former for parsing.

2. In a new cell, execute `graph = path_graph(3)` to create the graph we will save and load later.

3. Execute the following code in a new cell:

    ```
    savegraph("example.lg", graph, GraphMLFormat())
    ```

    In this case, we used the `savegraph` function described previously to save the graph in a file named `example.lg` using the *GraphML* format. Note that the format should be an instance of an `AbstractGraphFormat` subtype; this is `GraphMLFormat()` in this example.

4. Execute the following code in a new cell:

    ```
    loadgraph("example.lg", GraphMLFormat())
    ```

    The preceding code loads the graph stored in *GraphML* format from the file saved in the previous step.

If you use `MetaGraphs`, you can still use the `savegraph` and `loadgraph` functions. However, `MetaGraphs` uses the *MG format* by default to store the graph, along with the graph, edge, and vertex attributes. Also, the signatures from the `MetaGraphs` version of those functions are more straightforward than the signatures from `Graphs`. The `savegraph` function only takes the filename and the `MetaGraph` to save, while the `loadgraph` function takes the filename and an instance of `MGFormat`. For example, if you have a `MetaGraph` stored in the `metagraph.mg` file, you can read it by running `loadgraph("metagraph.mg", MGFormat())`. Here, we have used `MGFormat()` to create an instance of the `MGFormat` type.

Now that we have learned how to perform the most basic operations with graphs in Julia, let's learn how to visualize them.

Graph visualization packages

This section will showcase the basics of visualizing graphs in Julia using the `GraphPlot`, `GraphRecipes`, and `GraphMakie` packages. Other packages aim to plot graphs in Julia that we will not cover in this book; among those, there's `TikzGraphs`, `SankeyPlots`, and `EcologicalNetworksPlots`. Different packages have different strengths; let's focus on the three that we will explore in this chapter. First, we have the `GraphPlot` package, which has a straightforward interface. It has a single function, named `gplot`, with a few keyword arguments that are useful for quickly visualizing graphs. `GraphPlot`, similar to `Gadfly`, uses `Compose` to render the images. Therefore, it produces nice-looking bi-dimensional plots.

Then, we have the `GraphRecipes` package, which contains a collection of *Plots recipes*. You can benefit from the different `Plots` backends when working with this package. `GraphRecipes` also offers a function for visualizing graphs: `graphplot`. This function provides many keyword arguments that are very convenient when customizing your plot. Finally, the `GraphMakie` package offers a `graphplot` *Makie recipe* for plotting graphs. Like the function from `GraphRecipes`, it provides many ways to customize your plot. However, the main advantage of `GraphMakie` is its ability to add *interaction* to the plotted graph. Note that both `GraphRecipes` and `GraphMakie` can plot graphs in three dimensions.

There are two things that `GraphPlot`, `GraphRecipes`, and `GraphMakie` have in common that we should consider before starting. The first is that these packages use the term *node*, instead of the term *vertex*, which `Graphs` uses, in their documentation and keyword arguments. The second is that the three have keyword arguments to indicate the desired layout function to use. We will discuss that argument in detail in the *Graph layouts* section of this chapter.

Now that we have a general idea about the packages, let's start exploring them using the same graph. In this chapter, we will use *Zachary's karate club network*. It is a social network describing the interactions outside the classes, workouts, and club meetings among the university karate club members. The network contains 34 nodes or vertices and 78 edges. *Vertex 1* represents the karate instructor, *Mr. Hi*, while *vertex 34* represents the club administrator, *John A*; both are pseudonyms. A conflict between the two ended in the club's split, with some members forming a new club with Mr. Hi. Using the `Graphs` package, you can load this network by running `smallgraph(:karate)`. Let's start visualizing this network using the `GraphPlot` package.

Visualizing graphs with GraphPlot

The GraphPlot package exports the gplot function, which takes the graph to plot as the first positional argument. Optionally, it can take two more positional arguments indicating the vertex or *node positions*. For that, you should pass the positions using numeric vectors. The first vector determines the coordinates on the X-axis and the second the coordinates on the Y-axis. If we do not set those positional arguments, gplot will automatically select the vertex positions using the chosen layout function.

By default, gplot uses the *spring layout* algorithm. You can change the layout algorithm by passing a Symbol indicating the desired layout to the layout keyword argument. For example, you can use a *spectral layout* by setting the layout argument to :spectral_ layout. We will discuss the available layouts in the *Graph layouts* section of this chapter. One noticeable thing is that the GraphPlot package implements its layout algorithms rather than relying on other packages, which is was GraphRecipes and GraphMakie do.

The gplot function has the following keyword arguments to control the *node* or *vertex* properties. These keyword arguments can take a *single value* or a *vector* with as many values as there are nodes in the graph. Therefore, we can always pass a vector if we want to have different properties for different nodes. Let's see the arguments:

- nodesize: This argument determines the relative size of the nodes; therefore, you can pass a vector of values to make each node have a different size. Note that the vector should have positive values and that nodes with zeros are not displayed.

- nodefillc: This argument determines the color that fills the nodes. By default, they are "turquoise", but you can give any color using a *color string* that the Colors package can parse. For example, you can use a named color such as "gold", a hexadecimal color such as "#ffd700", or an RGB string such as "rgb(255,215,0)".

- Nodestrokec: This argument determines the color of the node stroke. Like nodefillc, it can take any string that the Colors package can parse. The gplot function sets this argument to nothing by default, avoiding the node strokes display. You will also need to change the node stroke's width to see the stroke.

- nodestrokelw: This argument sets the line width of the node stroke; by default, gplot sets this argument to 0.0, thus not drawing the stroke.

Similarly, there are two keyword arguments to customize the graph's *edges*. These arguments accept a *single value* or a *vector*. In the latter case, the vector should have as many values as there are edges in the graph:

- `edgestrokec`: This keyword argument determines the edge color, set to `"lightgray"` by default. As mentioned for the node color, you can select any color string that the `Colors` package can parse.

- `edgelinewidth`: This sets the relative line width of the edges. Note that the widths vector should only contain positive values.

Finally, the `gplot` function allows you to add *labels* for nodes and edges. You can perform this with the `nodelabel` and `edgelabel` keyword arguments, respectively. These arguments can take any vector or iterable containing objects that Julia can convert into strings. The node and edge label collection length should match the number of nodes and edges. Once you have node or edge labels, you can customize their size, color, and position using keyword arguments. Namely, `nodelabelsize` and `edgelabelsize` control the labels relative *font size*, while `nodelabelc` and `edgelabelc` control its *color*. These keyword arguments accept both single values and vectors. Then, the *position* of the node labels is determined using a circle centered in the same place as the node. Namely, we need to indicate the radius, the distance to the center, and the angle to define the point where we want to locate the node labels. The following keyword arguments set these positions while taking each node as a reference:

- `nodelabeldist`: This keyword argument takes a `Real` number specifying the distance from the node's center – the circle radius. By default, `gplot` sets this argument to zero, locating the labels on the node's center.

- `nodelabelangleoffset`: This keyword argument takes a `Real` number indicating an angle in the circle that determines the node label's location. For example, you can `nodelabeldist` to `2.5` to separate the label position from the node's center and set the angle offset to zero to position the label at the node's right. In the same way, setting the angle to `pi/2` will put the labels just over the nodes. By default, `gplots` places the labels in-between, using an angle offset of `pi/4`.

The `gplots` function locates the edge labels relative to the edge center. The distance from that point in the *X*-axis and *Y*-axis is determined using the `edgelabeldistx` and `edgelabeldisty` keyword arguments, respectively. By default, `gplot` sets both to zero while locating the edge labels in the edge's center.

Finally, if you plot a directed graph, gplot will automatically draw *arrowheads* on the edges to indicate the direction. You can control the *length* and *width* of the arrowheads using the arrowlengthfrac and arrowangleoffset keyword arguments, respectively.

Let's look at some examples to get a practical idea of how to work with the gplot function from GraphPlot using Pluto:

1. Create a new Pluto notebook and execute the following code in the first cell:

    ```
    using Graphs, GraphPlot
    ```

    Here, we have loaded the Graphs package to work with the graph and GraphPlot to visualize it.

2. Execute the following code in a new cell:

    ```
    graph = smallgraph(:karate)
    ```

    Here, we have loaded *Zachary's karate club network* into the graph variable.

3. In a new cell, execute the following code:

    ```
    gplot(graph)
    ```

    This is the most straightforward call to gplot as we are only giving the graph we want to plot and letting all the other arguments follow the default behavior. The following figure shows the plot that has been created:

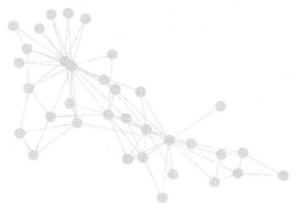

Figure 7.3 – Graph plotted with GraphPlot's gplot using default parameters

Note that the exact node positions could be different for you since the spring layout has a random component.

4. In a new cell, execute the following code:

```
gplot(graph,
    nodestrokelw = 1,
    nodestrokec = "lightgray")
```

Here, we have added the stroke line to the network's nodes. For that, we set its width to 1 using nodestrokelw and its color to "lightgray", the same color the edges use by default, using nodestrokec. Now, let's change the node fill color to highlight nodes 1 and 34.

5. Execute the following code in a new cell:

```
node_colors = fill("white", nv(graph))
```

Here, we have used Julia's fill function to create a new vector filled with the indicated value. We have made a vector with as many elements as there are vertices in the graph. We obtained that last value by calling the nv function from the Graphs package. We will color the unhighlighted nodes in white, so we have filled the vector of colors using the "white" string.

6. Execute the following code in a new cell:

```
begin
    node_colors[1] = "lightgray"
    node_colors[34] = "lightgray"
end
```

Here, we have changed the colors of nodes 1 and 34 from white to light gray. Suppose you are using Pluto's *dark mode*, or your monitor contrast doesn't allow you to differentiate this color. In that case, you can return to this cell and replace "lightgray" with another color, such as "orange". In any case, we recommend using *light mode* for this example to see the figure as it appears in this book.

7. Execute the following code in a new cell:

```
gplot(graph,
    nodestrokelw = 1,
    nodestrokec = "lightgray",
    nodefillc = node_colors)
```

Here, we passed the vector of colors that we created for the nodes to the nodefillc keyword argument to highlight those nodes using their color. Now, let's identify those nodes using labels.

8. In a new cell, execute the following code:

```
begin
    node_labels = fill("", nv(graph))
    node_labels[1] = "Mr. Hi"
    node_labels[34] = "John A"
end
```

This is similar to what we did when creating the vector of colors in *step 5* and *step 6*. We used the fill function to create a vector that has an empty string for each node. Then, we changed the label values of nodes 1 and 34 to identify *Mr. Hi* and *John A* in the network.

9. In a new cell, execute the following code:

```
gplot(graph,
    nodestrokelw = 1,
    nodestrokec = "lightgray",
    nodefillc = node_colors,
    nodelabel = node_labels,
    nodelabeldist = 2.5,
    nodelabelangleoffset = pi/2)
```

Here, we have added the labels to the node by passing the created vector of node labels to the nodelabel keyword argument. Then, we moved the labels up from the node centers using the nodelabeldist and nodelabelangleoffset keyword arguments. Note that GraphPlot shows the node labels in *black* by default, so if you use Pluto's *dark mode*, you will need to change their color to see them. You can color the labels orange by setting the nodelabelc attribute to GraphPlot.colorant"orange". The following figure shows the plot that has been created:

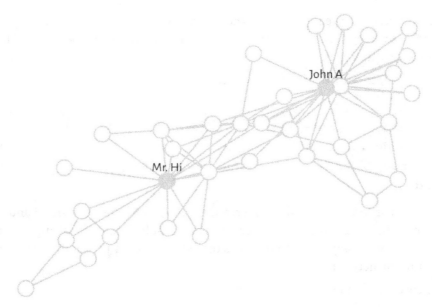

Figure 7.4 – Network with highlighted nodes using GraphPlot's gplot function

Now that we know how to visualize graphs using `GraphPlot`, let's look at `GraphRecipes`.

Exploring GraphRecipes

The `GraphRecipes` package offers a series of Plots recipes for graphs. The primary function to plot graphs with `GraphRecipes` is `graphplot`. This function can take different graph representations. Namely, you can input a *graph* from the `Graphs` package, an *adjacency matrix*, or an *adjacency list* represented by two vectors, one for the source node and the other for the destination node. We can use the adjacency matrix values to store *edge weights* or give a third vector of edge weights when using the adjacency list. This section uses the term node since `GraphRecipes` refers to *vertices* as *nodes*.

The `graphplot` function from `GraphRecipes` can take the attributes from `Plots` when possible. The complete list of `Plots` arguments will be described in *Chapter 12, Customizing Plot Attributes – Axes, Legends, and Colors*. This section will only focus on the keyword arguments from `graphplot` that are related uniquely to graph visualization. When available, we will also mention the attribute aliases.

Let's start describing the keyword arguments that we can use to modify the different aspects of *nodes*:

- `nodeshape`, `node_shape`, or `markershape`: This argument determines the geometric shape of the nodes. It takes a symbol or a vector of symbols to assign different shapes to different nodes. The available symbols are `:hexagon`, `:circle`, `:ellipse`, and `:rect`. By default, `GraphRecipes` uses `:hexagon`. You can also pass a function to this keyword argument to create custom shapes for the nodes. We will see an example later in this section.

- `nodecolor`, `marker_color`, or `markercolor`: This argument determines the color of the nodes. You can pass any color that the `Plots` package can use. You can also give a vector of colors to have each node colored differently. If you set this argument using an integer, `GraphRecipes` will take the color in that position from the current palette. By default, this argument is set to `1` to take the first color from the color palette.

- `nodesize`, `node_size`, or `markersize`: This keyword argument takes a floating-point number to determine the nodes size. By default, `graphplot` sets this attribute to `0.1`. If we label the nodes using the `names` attribute, `GraphRecipes` will scale the nodes to fit the labels inside them.

- `node_weights` or `nodeweights`: This argument allows us to scale the node sizes according to a vector of numerical values; the weights.

- `names`: This attribute labels the nodes. You can pass a vector or iterable of objects that Julia can convert into strings. `GraphRecipes` will scale the nodes to fit the labels; therefore, be careful when assigning labels with different sizes. We will see an example of this later in this section.

`GraphRecipes` draws nodes using a scatter plot; therefore, you can use *marker attributes* with `graphplot`. You can use those arguments or their aliases. To get the `graphplot` keyword argument, you only need to change the *marker* prefix of the attribute via the *node* prefix. For example, you can change the node's transparency using the `markeralpha` or `nodealpha` attribute.

`GraphRecipes` also offers a series of keyword arguments for customizing the graph's *edges*. This package identifies the edges using their source and destination nodes; you can get them using the `src` and `dst` functions from the `Graphs` package. The `graphplot` function from `GraphRecipes` offers the following keyword arguments for customizing the edges:

- `curves`: This is a Boolean argument that controls whether `graphplot` will draw the edges as curves, which is the default behavior, or as straight lines. The curves are *cubic splines* that go from the source node to the destination node, passing through a point at a given distance from the average node positions. The `curvature_scalar` keyword argument controls the distance from that point.

- `curvature_scalar`, `curvaturescalar`, or `curvature`: This argument controls the distance of the point that defines the trajectory of the curved edge from the middle of the straight line linking the two nodes. Therefore, setting `curvature_scalar` to zero will draw straight lines instead of curves. By default, `graphplot` sets this value to `0.05`.

- `shorten` or `shorten_edge`: This keyword argument controls how much to trim from the edge extremes. Therefore, it takes a floating-point number between `0.0` and `0.5` to indicate the edge fraction that's cut from each extremum. The `graphplot` function sets this value to `0.0` by default to avoid shortening the edges.

- `edgewidth`, `edge_width`, or `ew`: This keyword argument determines the edge widths, considering the source and destination nodes and the edge weights. This attribute can take a function, a dictionary, or a matrix. The *function* should take three arguments – the source and destination nodes and the edge weight – and return the line width for that edge. For example, `graphplot` sets this argument to `(src, dst, weight) -> 1` by default to give the same width to all the edges. The *dictionary* should have source and destination node tuples as keys and line widths as values. We will see an example of this later in this section. If we provide a *matrix*, `graphplot` will index the matrix using the node numbers; therefore, the matrix should store the line widths.

- `edgelabel`, `edge_label`, or `el`: This keyword argument allows you to set labels for the edges. It can take a vector, matrix, or dictionary to define the edge labels. If we use a dictionary, its keys should be a tuple of the source and destination node numbers. The labels can be any object that Julia can convert into a string; however, note that `graphplot` will not display the labels that are empty strings, `missing`, `false`, `nothing`, or `NaN`.

- `edgelabel_offset`, `edgelabeloffset`, or `elo`: This attribute determines the distance between the edge and its label. The `graphplot` function sets this argument to `0.0` by default.

- `edge_label_box`, `edgelabelbox`, `edgelabel_box`, or `elb`: This Boolean argument determines whether the labels should have a box to avoid the intersection between the text and edge lines. By default, `graphplot` does not draw the boxes around the labels.

- `self_edge_size`, `selfedgesize`, or `ses`: This keyword argument determines the size of self-edges; that is, edges linking a node to itself. The `graphplot` function sets this attribute to `0.1` by default.

Finally, the `graphplot` function also offers a series of attributes to control some general aspects of the plot. We can use the `method`, `func`, `layout_kw`, `x`, `y`, and `z` attributes to set up the layout algorithm and determine the nodes' positions. We will discuss these arguments in the *Graph layout* section. Some other practical keyword arguments are as follows:

- `dim`: This determines the number of dimensions to use for the visualization. By default, `graphplot` sets this attribute to 2 to plot the graphs in a plane. We can set this argument to 3 to have *tridimensional graph plots*. We recommend using an interactive backend such as *Plotly* for the tridimensional graphs as we can easily rotate the figure to avoid the occlusion problem.

- `root`: When displaying *trees*, you can indicate `graphplot` to position the tree root at the `:top` (the default), `:bottom`, `:left`, or `:right` area of the figure.

- `fontsize`: This attribute controls the font size of the node and edge labels. By default, `graphplot` uses a font size of 7 points.

- `axis_buffer` or `axisbuffer`: We can use this argument to increase or decrease the margin around the graph. By using the fraction, as indicated by this argument, `graphplot` increases the *x*, *y*, and *z* limits. By default, there is an increase of `0.2` in each direction; you can increase this number if the graph goes outside the plot limits.

Now that we have discussed the main keyword arguments of the `graphplot` function, let's explore some of them using *Pluto*:

1. Execute the following code in the first cell:

   ```
   using Graphs, Plots, GraphRecipes
   ```

   Here, we have loaded `Graphs` to work with the graph and `Plots` and `GraphRecipes` to explore the last package.

2. In a new cell, execute `graph = smallgraph(:karate)` to load *Zachary's karate club network*.

3. Execute the following code in a new cell:

   ```
   graphplot(graph)
   ```

   Here, we have created the first plot of the network. As shown in *Figure 7.1*, using `graphplot` from `GraphRecipes` with default parameters produces a graph with curved edges and hexagonal nodes.

4. Execute the following code in a new cell:

   ```
   graphplot(graph,
        shorten = 0.15,
        curves = false)
   ```

Here, we have set the `curves` keyword argument to `false` to draw edges as straight lines instead of curves. We have also set the `shorten` attribute to `0.15` to cut the edge extremes near the nodes. The following figure shows the plot that has been created:

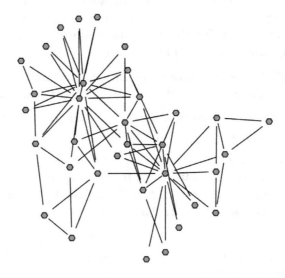

Figure 7.5 – A graph with straight and trimmed edges

5. In a new cell, execute the following code:

```
begin
    colors = fill(:white, nv(graph))
    colors[1] = :darkgray
    colors[34] = :darkgray
end
```

Here, we have created a vector of colors, using symbols for their names, using Julia's `fill` function. The constructed vector has a length identical to the number of vertices or nodes in our graph. All the values are `:white`, so we assigned `:darkgray` to nodes 1 and 34 to highlight them.

6. In a new cell, execute the following code:

```
graphplot(graph,
    nodecolor = colors,
    nodestrokecolor = :darkgray,
    edgecolor = :darkgray,
    nodeshape = :circle,
```

```
nodesize = 0.2,
curves = false)
```

Here, we have set the nodecolor attribute using the colors vector that was created in the previous step. Then, we changed the line colors for the nodes and edges to :darkgray using the nodestrokecolor and edgecolor attributes, respectively. We used the nodeshape argument to plot the nodes using circles instead of hexagons and used nodesize to make them bigger. Finally, we set curves to false to draw the edges as straight lines. The following figure shows the plot that has been created:

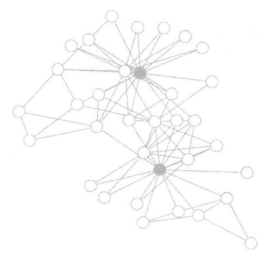

Figure 7.6 – A graph with nodes highlighted using colors

Now, let's draw nodes using custom shapes.

7. In a new cell, execute the following code:

```
function customshape(x, y, nodeheight, nodewidth)
    [
    (x - nodewidth/2, y - nodeheight/3),
    (x + nodewidth/2, y - nodeheight/3),
    (x, y + 2nodeheight/3)
    ]
end
```

Here, we have defined a function named customshape that takes four positional arguments. The x and y arguments will determine where we will center the shape, while the `nodeheight` and `nodewidth` arguments will indicate the desired height and width for the node. The function should return a vector of tuples, in which each tuple contains the *x* and *y* coordinates of a shape's vertex. In this case, we produce a vector of three points to draw a triangular shape. In particular, each point indicates the vertex coordinates for an isosceles triangle with its centroid on the *(x,y)* point.

8. In a new cell, execute the following code:

```
function customshape(x, y, nodescale)
    [
    (x - nodescale/2, y - nodescale/3),
    (x + nodescale/2, y - nodescale/3),
    (x, y + 2nodescale/3)
    ]
end
```

For symmetrically scaling shapes, such as our isosceles triangle, we can optionally define a function that takes three arguments instead of four. Again, the x and y arguments determine the shape's center. The third argument, `nodescale`, determines the size of our node. As with the previous step, the function returns a vector with three tuples. Each tuple has the vertex coordinates for an isosceles triangle centered in *(x, y)*.

9. Run the following code in a new cell:

```
graphplot(graph,
    node_shape = customshape,
    nodecolor = :white,
    nodesize = 0.2,
    curvature_scalar = 0.01)
```

Here, we have provided the `node_shape` attribute with the function we defined in the two previous steps. As you can see, our nodes are now triangles. We have also changed the color of the nodes using the `nodecolor` attribute, this time using a single color for all the nodes. We have set `nodesize` to `0.2` to make the nodes bigger than the default. Finally, to make the edge curvature more subtle, we have changed the `curvature_scalar` attribute from `0.05`, which is the default, to `0.01`. The following figure shows the plot that has been created:

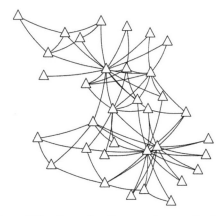

Figure 7.7 – A graph using a custom node shape

10. Run the following code in a new cell:

```
begin
    light_highlight = fill(:white, nv(graph))
    light_highlight[1] = :lightgray
    light_highlight[34] = :lightgray
end
```

The preceding code is similar to the code we executed in *step 5*. The main difference is that we used :lightgray instead of :darkgray to highlight nodes 1 and 34 since we will add labels inside the nodes in the following steps.

11. In a new cell, execute the following code:

```
different_lengths = graphplot(graph,
    names = 1:nv(graph),
    nodecolor = light_highlight,
    nodeshape = :rect,
    nodesize = 0.2,
    curves = false,
    axis_buffer = 0.1)
```

Here, we have plotted our graph, adding node labels with the names attribute. The node labels are simply the node numbers, from 1 to the number of vertices in the graph. We have also assigned :rect to the nodeshape attribute to get rectangular nodes. We have also colored the nodes using the light_highlight variable we created in the previous step. We have customized the other node and edge attributes by doing what we did previously; we used nodes larger than the default and straight edges. We also reduced the margin around the graph using the `axis_buffer` attribute.

Looking at the graph and looking for nodes 1 and 34, you will notice that the nodes have different sizes. That's because the node sizes are scaled to fit the labels, and node 1 has a label that is one character shorter than the label for node 34, as shown on the left in *Figure 7.8*. To avoid node size differences, we can create all the labels with the same length; let's try that.

12. Run `using Printf` in a new cell to load the module that's needed to format the strings.

13. In a new cell, run the following code:

```
node_labels = [@sprintf "%2d" i for i in 1:nv(graph)]
```

Here, we have used the `@sprintf` macro to interpolate the node number into a string, ensuring that the number has a width of 2. We created the vector of labels using an array comprehension and iterated on the range of node numbers. You can see from the output that the string for number one is " 1" instead of "1".

14. Execute the following code in a new cell:

```
equal_lengths = graphplot(graph,
    names = node_labels,
    nodecolor = light_highlight,
    nodeshape = :rect,
    nodesize = 0.2,
    curves = false,
    axis_buffer = 0.1)
```

Here, we have plotted our graph using the same arguments as in *step 11*, except that we have set the `names` attribute with the variable we defined in the previous step. Since all the labels have the same number of characters, all the nodes are the same size. Let's plot the two graphs together so that we can compare them.

15. Execute `plot(different_lengths, equal_lengths)` in a new cell. The following figure shows the plot that has been created:

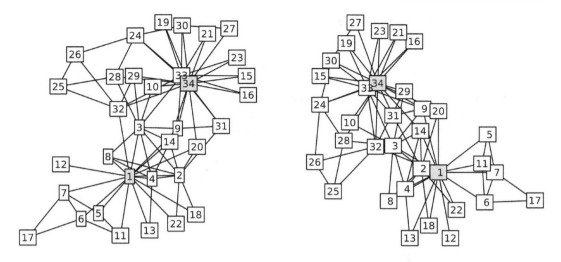

Figure 7.8 – Graphs with different (left) and equal (right) length labels

16. In a new cell, execute the following code:

```
edge_widths = Dict(
    (src(edge), dst(edge)) => 0.1 + i/ne(graph)
    for (i, edge) in enumerate(edges(graph)))
```

Here, we have created a dictionary that goes from a node tuple, defining the edges, to the desired edge width. We have used an iterator syntax similar to array comprehension to make the dictionary. The `enumerate` iterator goes through an iterator – in this case, the one returned by the `edges` function – and returns a tuple. The tuple has the iteration number as the first element and the element produced by the input iterator as the second. We took advantage of Julia's destructuring assignment to assign the first element to the `i` variable and the second to the `edge` variable.

To populate the dictionary, we produced a `Pair`, using the `=>` operator, on each iteration. The first element of the pair is the key, while the second is the value. Here, the key is a tuple with the edge's source node in the first position and the destination node in the second. We have accessed the source and destination nodes of the edge using the `src` and `dst` functions from `Graphs`, respectively. For the value, we set a minimum width of `0.1`, and then we added a value that is proportional to the edge number, `i`. We divided `i` by the number of edges to avoid going beyond 1.1 and creating enormous edges.

17. In a new cell, execute the following code:

```
graphplot(graph,
    edgewidth = edge_widths)
```

Here, we have plotted our graph using different widths for our edges by assigning the dictionary created in the previous step to the edgewidth attribute. The following figure shows the plot that has been created:

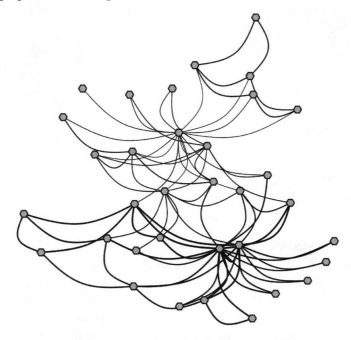

Figure 7.9 – A graph with different edge widths

18. Execute the following code block in a new cell:

```
code = :(
    while condition
    do_something
    end
)
```

Here, we have assigned a Julia expression, a while loop, of the Expr type to the code variable. We used the :(and) syntax to create the quoted expression.

19. Execute the following code block in a new cell:

```
plot(code,
    nodecolor = :white,
    fontsize = 10,
    nodeshape = :rect,
    axis_buffer = 0.4,
    root = :left)
```

Here, we have called the `plot` function from the `Plots` package to plot the `Expr` object in the `code` variable. `GraphRecipes` has a dedicated recipe for the `Expr` object that plots the **Abstract Syntax Tree** (**AST**). This recipe relies on `graphplot`; therefore, we can use the previously described attributes. For example, we have used `nodeshape` to plot the nodes as rectangles and `axis_buffer` to avoid part of the nodes going outside the plot. We have also used the `root` attribute to have the tree's root on the left and the leaves on the right. The following figure shows the plot that has been created:

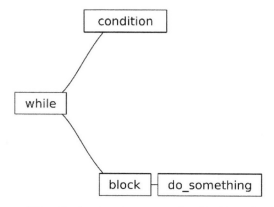

Figure 7.10 – Abstract syntax tree made with GraphRecipes

20. In a new cell, execute the following code:

```
plot(Graphs.AbstractSimpleGraph,
    nodecolor = :white,
    nodestrokewidth = 0.0,
    nodeshape = :rect)
```

Here, we have used the recipe that `GraphRecipes` provides to plot the inheritance tree of Julia's types using the `plot` function. Here, we plotted the type hierarchy tree for `AbstractSimpleGraph`. As we did previously, we can use the attributes described for `graphplot` with this recipe. In this example, we have drawn white nodes without borders by setting `nodecolor` to `:white` and `nodestrokewidth` to `0.0`. The following figure shows the plot that has been created:

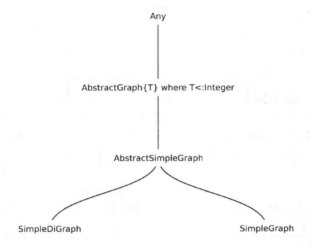

Figure 7.11 – Inheritance tree made with GraphRecipes

In this section, we learned how to visualize graphs using the `graphplot` function from the `GraphRecipes` package. We explored the main attributes available for customizing those graphs, and we saw examples for some of them. Lastly, we saw the package's recipes for visualizing abstract syntax trees and type hierarchies.

Using `GraphRecipes`, you can benefit from the interactivity provided by the different `Plots` backends. For example, you can use the *Plotly* backend to allow *panning* and *zooming*. However, `Plots` backends do not allow you to add interactivity to nodes and edges. In the next section, we will quickly explore the best package in the Julia ecosystem for creating interactive graph visualizations: `GraphMakie`.

Creating interactive graphs with GraphMakie

The `GraphMakie` package offers the `graphplot` *Makie recipe* for visualizing graphs. It also provides a series of predefined interactions for moving and highlighting nodes and edges. As with the previously described visualization packages, and unlike `Graphs`, `GraphMakie` uses the *node* term to name *vertices*. The graph comprises a set of *layers*, a `scatter` plot to draw the *nodes*, a `text` plot for the *node labels*, and another for the *edge labels*. If we plot a directed graph, there is another `scatter` plot to draw the *arrows*. Finally, the *edges* layer is an `edgeplot` figure. The `edgeplot` recipe from `GraphMakie` relies on `linesegments` for straight edges and `lines` for curvy edges.

The `graphplot` function exposes the main attributes for the node, edge, and label layers through keyword arguments. Most of those arguments can take a single element or a vector with as many elements as nodes or edges are in the graph. The function also offers the `node_attr`, `edge_attr`, `arrow_attr`, `nlabels_attr`, and `elabels_attr` keyword arguments to pass extra arguments into the corresponding layers. Those attributes with the `attr` suffix take a named tuple as input. A full description of the keyword arguments of the `scatter`, `lines`, `linesegments`, and `text` functions has been provided in *Chapter 12, Customizing Plot Attributes – Axes, Legends, and Colors.*

Let's see the keyword arguments that `graphplot` exposes to customize our graph visualizations. The `graphplot` function passes the values of the `node_color`, `node_marker`, and `node_size` arguments to the `color`, `marker`, and `markersize` arguments of the `scatter` function to customize the *nodes*.

For customizing *edges*, the most crucial keyword argument is `edge_plottype`. You can set that argument to `:linesegments` to have straight edges or `:beziersegments` to have curved edges. Note that choosing the latter can be slow for big graphs. `GraphMakie` decides the edge type by default, depending on the current edge number and types. Then, `graphplot` passes the values of the `edge_color` and `edge_width` arguments to the `color` and `linewidth` arguments of the `lines` or `linesegments` function.

`GraphMakie` controls the *arrowheads* using the following keyword arguments. The `arrow_show` takes a Boolean argument that `graphplot` automatically sets to `true` for directed graphs; as its name suggests, it decides whether to show the arrowheads. Then, `graphplot` will pass the value of the `arrow_size` argument to the `markersize` argument of the corresponding `scatter` plot. Finally, the `arrow_shift` attribute determines where to draw the arrowhead. It takes a floating-point number between `0.0`, locating the arrowhead at the source node, and `1.0`, locating it at the destination. By default, `graphplot` sets `arrow_shift` to `0.5`, locating the arrowhead at the edge center.

The `GraphMakie` package offers similar keyword arguments for setting and customizing *node and edge labels*; the former uses the `nlabels` prefix, while the latter uses the `elabels` prefix. For example, we must pass a vector of strings to `nlabels` to set node labels or `elabels` to set edge labels. The `graphplot` function gives the values of the `nlabels_align`, `nlabels_color`, and `nlabels_textsize` attributes – or `elabels_align`, `elabels_color`, and `elabels_textsize` for edge labels – to the `align`, `color`, and `textsize` arguments of the `text` function, respectively.

Since the label positions are set differently for node and edge labels, the keyword argument to set those positions also differs. Let's start with *node labels*; their positions are relative to the node's position. We can use `nlabels_distance` to indicate the distance in pixels between the point textbox's anchor and the node. The `nlabels_align` attribute determines the anchor position in the textbox; therefore, it determines the direction in which graphplot will place the label. If we increase nlabels_distance, graphplot will push the label's textbox away from the node, following a direction opposite what's indicated by nlabels_align. For example, since nlabels_align is (:left, :bottom) by default, increasing nlabels_distance will move the label in the upper right direction.

Sometimes, it can be convenient to move the label's reference position from the node to another point. The graphplot function offers the nlabels_offset attribute to achieve this. This argument takes a Point or a vector of Point objects in the data space that graphplot adds to the node's positions to move the reference position. If we do that, the nlabels_distance and nlabels_align attributes will use the new points, rather than the nodes, as the centers.

The graphplot function locates an *edge label* at a given fraction of an edge, measured from the source node. We can set that fraction using the elabels_shift keyword argument, which works similarly to the previously described arrow_shift. By default, GraphMakie sets elabels_shift to 0.5, locating the labels at the middle of the edge. Then, we can use elabels_distance and elabels_align, similarly to the previously described nlabels_distance and nlabels_align, to place the labels relative to their position on the edge. One difference is that graphplot sets elabels_align to (:center, :bottom) by default.

There is also an elabels_offset attribute, similar to nlabels_offset, to move the reference point. The edge labels have two special attributes we can set: elabels_rotation and elabels_opposite. The former takes an angle for each label to define the text rotation; by default, graphplot decides the rotations based on the edge angle. The latter takes a list of edge indices to indicate the edges that display their labels on the opposite side.

Finally, the graphplot function from GraphMakie offers a series of arguments to customize self and curved edges, which we will not discuss in this book. Among those arguments, you will find the curve_distance attribute, which graphplot sets to 0.1 by default. It works similarly to the previously described curvature_scalar attribute from GraphRecipes; one crucial difference is that curve_distance can also take a vector of distances or a dictionary to assign different curvatures to different edges.

Now that we know the main argument for customizing the aspect of our graph plot using GraphMakie, let's explore some of them and the predefined interactions using the following example:

1. Create a new Pluto notebook and execute the following code in the first cell:

    ```
    begin
    ```

```
using Graphs
    using GLMakie, GraphMakie
    end
```

Here, we have loaded the `Graphs` package to work with the graph and `GraphMakie` to visualize it. We have chosen the `GLMakie` backend to create an interactive graph. Please wait as this cell can take a few minutes to execute for the first time.

2. Execute `graph = smallgraph(:karate)` in a new cell to load *Zachary's karate club network*.

3. Execute `graphplot(graph)` to create the most straightforward graph plot possible with `GraphMakie`. Note that, by default, `GraphMakie` shows the coordinates system and displays the nodes as points and the edges as straight lines. The plot can take some time to appear. The following figure shows the plot that has been created:

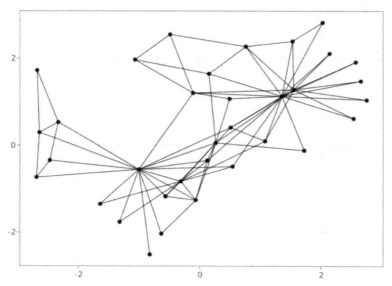

Figure 7.12 – Plot created with graphplot from GraphMakie with default parameters

Let's customize this visualization to make it similar to the one shown in *Figure 7.4*.

4. In a new cell, execute the following code:

```
begin
        node_colors = fill(:white, nv(graph))
        node_colors[1] = :lightgray
```

```
        node_colors[34] = :lightgray
    end
```

Here, we have created a vector of colors with as many :white symbols as there are vertices or nodes in the graph. Then, we changed the color name at positions 1 and 34 to :lightgray to highlight those nodes later.

5. In a new cell, execute the following code:

```
begin
        node_labels = fill("", nv(graph))
        node_labels[1] = "Mr. Hi"
        node_labels[34] = "John A"
    end
```

Here, we have created a vector of node labels. Since we only want to add labels to nodes 1 and 34, we created a vector of length 34 – the number of vertices in our graph – that contains empty strings. Then, we assigned the pseudonyms of the individuals in nodes 1 and 34 to the node_labels vector.

6. Execute the following code in a new cell:

```
begin
    figure, axis, plt = graphplot(graph,
        node_color = node_colors,
        edge_color = :lightgray,
        node_size = fill(16, nv(graph)),
        edge_width = fill(2.5, ne(graph)),
        node_attr = (strokewidth = 2.5,
            strokecolor = :lightgray),
        nlabels = node_labels,
        nlabels_align = (:center, :bottom),
        nlabels_distance = 10)
    hidespines!(axis)
    hidedecorations!(axis)
    figure
end
```

Here, we used the previously described keyword argument of the graphplot function from GraphMakie to replicate the plot in *Figure 7.4*. As we did previously, we passed the graph to the graphplot function. Like other Makie plotting functions, that function returns a tuple containing the `Figure`, `Axis`, and `Plot` objects. We used Julia's destructuring assignment to assign those objects to the `figure`, `axis`, and `plt` variables, respectively, as we will need them later.

Let's explain how we have customized our plot using different `graphplot` keyword arguments. First, we set the node color by passing the vector created in *step 4* to `node_color`. Then, we set the edge colors to `:lightgray` using the `edge_color` keyword argument.

We determined the size of the nodes using the `node_size` argument. Interestingly, we have given a vector of length 34 – the number of vertices in our graph – and `fill` with a value of 16. While we can write `node_size = 16` to get the same output, we need the `node_size` and `edge_width` arguments to take a *vector* to add the `GraphMakie` *predefined interactions* in the following steps. That's why we set `edge_width` to a vector that has as many 2.5 values as there are edges in the graph.

To customize the node aspect further, we used the `node_attr` keyword argument to pass attributes to the `scatter` plot function that draws the nodes. We gave a named tuple to set the `strokewidth` and `strokecolor` attributes of the points.

We used `nlabels` to set the node labels using the vector of labels that was created in *step 5*. We located the anchor at the bottom center of the textbox using the `nlabels_align` keyword arguments. Finally, we moved the textbox 10 pixels away from the node using `nlabels_distance`; the `graphplot` function pushes the labels up because the anchor is at the bottom.

The created figure will show the **axis spines** – the lines delimiting the data space – and all the axis decorations, as shown in *Figure 7.12*. Therefore, to get a clean figure like the one created by `GraphPlot`, we called the `hidespines!` and `hidedecorations!` functions on the `Axis` object to hide the axis spines and decorations, respectively.

Then, we let the cell return the `Figure` object to visualize the created graph plot in the Pluto notebook. We can see that the produced plot is similar to the one shown in *Figure 7.4*. Now, let's add the predefined graph interactions available in `GraphMakie`:

1. In a new cell, execute the following code:

```
begin
    deregister_interaction!(axis, :rectanglezoom)
    register_interaction!(axis, :ndrag, NodeDrag(plt))
    register_interaction!(axis, :edrag, EdgeDrag(plt))
    register_interaction!(axis, :nhover,
        NodeHoverHighlight(plt))
```

```
        register_interaction!(axis, :ehover,
            EdgeHoverHighlight(plt))
end
```

First, we called the deregister_interaction! function to deregister the default :rectanglezoom interaction from the Axis object, as needed for the *drag and drop interactions*. Then, we added the predefined node and edge drag interactions to the Axis object using the register_interaction! function and the NodeDrag and EdgeDrag functions, respectively. Those functions take the plot object created by graphplot.

Finally, we add the predefined *highlight interactions* defined by the NodeHoverHighlight and EdgeHoverHighlight functions. These interactions increase the size of the nodes and edges on mouse hover events. Both functions take a factor keyword argument to determine the magnification factor used; it is 2 by default. These functions need the node_width, edge_width, and, optionally, arrow_width attributes to be vectors.

Besides these predefined interactions, GraphMakie also offers a set of helper functions to add user-defined behaviors upon *hovering over*, *clicking*, and *dragging* events for nodes and edges. We won't cover all these elements here, but let's set an action for node click events as an example before displaying our interactive graph.

2. Execute the following code in a new cell:

```
function node_action(idx, event, axis)
    if plt.node_color[][idx] == :white
        plt.node_color[][idx] = :lightgray
    else
        plt.node_color[][idx] = :white
    end
    plt.node_color[] = plt.node_color[]
end
```

Here, we have defined a function that takes three arguments; the first, idx, is the index of the clicked node. This function has side effects as it internally modifies the plot object; plt. In the function's body, we accessed the node_color attribute of the plot object. This attribute is an Observable object, which we explained in *Chapter 3, Getting Interactive Plots with Julia*. Here, we switched the node_color in the idx position between :white and :lightgray. The last line of the function triggers a change event on the observable object.

3. Execute the following code in a new cell:

```
register_interaction!(axis, :nodeclick,
    NodeClickHandler(node_action))
```

Here, we have used the register_interaction! function to add an interaction named :nodeclick to the Axis object. We created the interaction using the NodeClickHandler function, which takes the function that is executed on node click events. The function should take three positional arguments; in this case, we used the one defined in the previous step.

4. Execute display(figure) in a new cell. This will open a GLFW window showing our interacting graph, as shown here:

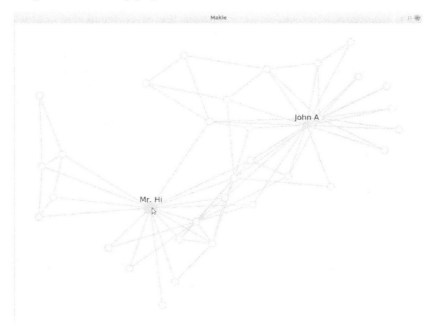

Figure 7.13 – GLFW window showing the interactive graph

Here, the mouse pointer was over the node corresponding to *Mr. Hi*, where the node is bigger than the other.

With that, we've learned how to create interactive graph visualizations using `GraphMakie`. We have seen the essential keyword arguments of the `graphplot` function from `GraphMakie` so that we can start customizing our graph plots. In this section, we learned how to draw graphs using the Julia ecosystem's three main libraries: `GraphPlot`, `GraphRecipes`, and `GraphMakie`. In the next section, we will learn how to set different layouts using those libraries.

Graph layouts

Graph layout algorithms determine the location of nodes and edges to create a readable visual representation of the graph. Readable graphs usually have few links crossing, large angles between the crossing links, and few node superpositions. When using the `gplot` function from `GraphPlot`, or the `graphplot` function from `GraphRecipes` and `GraphMakie`, we can select different graph layout algorithms or directly input the nodes and edge positions. In this section, we will explore both options. We will also introduce the `NetworkLayout` package as it implements several graph layout algorithms that we can use with the three libraries mentioned earlier.

Setting node positions can help us use external layout methods. It can also help when working with networks whose nodes have a known spatial location; for example, think of a subway network. However, note that using spatial locations can lead to graphs that can't be read easily. To set the node's exact location using `gplot` from `GraphPlot`, you need to call the function using three positional arguments; a graph, a vector with the nodes' coordinates on the *X*-axis, and one with the *Y*-axis coordinates. `GraphPlot` implements its layout algorithms; therefore, their layout functions are adapted to the `gplot` signature and return a tuple of *X*-axis and *Y*-axis coordinates vectors. The `graphplot` function from the `GraphRecipes` package allows you to determine the node positions using the x, y, and z keyword arguments. These arguments take an iterable that returns the coordinates in the corresponding axis.

The `graphplot` function from the `GraphMakie` package doesn't have a keyword argument to pass the node coordinates directly. However, we can pass a user-defined function to its `layout` attribute. That function should take the graph as input and output the node coordinates as a vector of `Point` objects. Therefore, we can create a function that returns a vector with our desired node coordinates. For example, suppose we have an xs vector holding the *X*-axis coordinates of the nodes and a ys vector for the positions on the *Y*-axis. In that case, we can pass an anonymous function such as `_ -> Point.(zip(xs, ys))` to the `layout` keyword argument.

Another nice `GraphMakie` feature is that the node's positions are stored as an `Observable` object that we can access using the `node_pos` attribute of the plot object. For example, if we have a `plt` variable holding the plot object, we can locate the first node at the origin by running the following code:

```
plt.node_pos[][1] = Point(0, 0)
notify(plt.node_pos)
```

We call the `notify` function to trigger the plot update after modifying the coordinates of the node. Since `node_pos` holds a vector of `Point` objects, we assigned an object of that type.

The `graphplot` function from `GraphMakie` allows us to determine edge paths using the `waypoints` and `waypoint_radius` keyword arguments. This is particularly useful for layout algorithms that determine the edge paths, such as the `Zarate` algorithm from the `LayeredLayouts` package.

Most of the layout algorithms in the `GraphRecipes` package depend on the `NetworkLayout` package. This package offers a series of types for setting different layout algorithms using keyword arguments. When we call the constructor, it returns a callable object. That object takes a graph or an adjacency matrix and returns a vector of `Point` objects. Therefore, we can feed the `layout` keyword argument of *GraphMakie's* `graphplot` function with those objects.

The `graphplot` function from `GraphRecipes` wraps some layouts from that library; we can select them using the `method` keyword argument. That argument takes a `Symbol` with the name of the layout method to execute. If we need to pass keyword arguments to the layout constructor, we can provide them to the `layout_kw` attribute of `graphplot` as a dictionary that goes from the keyword argument names as symbols to the desired values. If needed, `graphplot` also has a low-level `func` keyword argument that takes a function that returns the node's positions.

Let's see the main graph layout algorithms and how we can use them in `GraphPlot`, `GraphRecipes`, and `GraphMakie`:

- **Spectral layout**: This algorithm uses the eigenvectors of the Laplacian graph to determine the node's positions. It is not very popular, but it can be faster than the more popular string-based methods. You can pass the `spectral_layout` function to the `layout` keyword argument of the `gplot` function of `GraphPlot` to use this algorithm with that package. For example, if we have our graph stored in the `graph` variable, the call would look like this:

  ```
  gplot(graph, layout=spectral_layout)
  ```

The `NetworkLayout` package implements this algorithm on the `Spectral` type. The type has two keyword arguments; `dim` to determine the spatial dimensions for the output points and the node positions, and `nodeweights` to pass a vector of node weights. The `dim` argument is 3 by default, so we need to set `dim` to 2 to create a bi-dimensional plot. For example, to visualize a graph in two dimensions using `GraphMakie` and this layout method, we can do the following:

```
figure, axis, plt = graphplot(graph,
    layout=Spectral(dim=2))
```

The `graphplot` function from `GraphRecipes` also uses the spectral layout implementation of `NetworkLayout`. To use it, we need to set the `method` keyword argument using the `:spectral` symbol, like so:

```
graphplot(graph, method=:spectral)
```

- **Spring layout**: This uses an attractive force between neighboring nodes and a repulsive force between all nodes. This allows us to have the pairs of connected nodes as close as possible while preventing them from collapsing. `GraphPlot` uses this algorithm by default through its `spring_layout` function. The `NetworkLayout` package implements this algorithm on its `Spring` type. *GraphMakie's* `graphplot` uses that `Spring` type by default. *GraphRecipes'* `graphplot` also uses the `NetworkLayout` implementation when we set `method` to `:spring`.

- **Stress majorization**: This algorithm minimizes an energy or stress function to determine the node positions. It is the layout algorithm that's used by default by `GraphRecipes` since it has its own implementation. `GraphMakie` can use the `Stress` type from `NetworkLayout`. `GraphPlot` implements this method in the `stressmajorize_layout` function.

- **Circular and shell layouts**: These layout algorithms place the node in circles. The circular layout places them in a circle, while the shell layout places them in concentric circles – that is, shells. In practice, a shell layout with a single shell is a circular layout. The `NetworkLayout` package implements these layouts in the `Shell` type. This type takes a `nlist` keyword argument that takes a vector of shell indices, from the inner shell to the outer shell. If we do not give a list, it will create a single shell. We can pass that type to the layout argument of the `graphplot` function from `GraphMakie`.

When using the `GraphRecipes` package, we can set the `method` argument to `:circular` or `:shell` to use the `NetworkLayout` implementation. The `GraphPlot` package offers its implementations on the `circular_layout` and `shell_layout` functions. `shell_layout` takes the graph or adjacency matrix in its first positional argument and, optionally, a list of shell indices for each node on its second argument.

The `NetworkLayout` also has a **Scalable Force Directed Placement (SFDP)** layout implemented in the `SFDP` type. This can be accessed by `GraphRecipes` through the `:sfdp` method. For *visualizing trees*, the `NetworkLayout` package offers the `Buchheim` type, which implements **Buchheim's Tree Drawing Algorithm**. You can access it from `GraphRecipes` using the `:buchheim` method. The `GraphRecipes` package also provides the `:tree` method for drawing trees.

Finally, `NetworkLayout` exports a `SquareGrid` type to locate the nodes in a bi-dimensional grid. For graphs where the node ordering is meaningful, `GraphRecipes` offers the `:arcdiagram` method to draw *arc diagrams* and the `:chorddiagram` method for *chord diagrams*.

This section has taught us how to set the node's positions using different graph layout algorithms. This knowledge will help us choose the best layout for our visualizations. In the next section, we will briefly mention the main analysis tools that are available in the `Graphs` library.

Analyzing graphs

In the previous sections, we learned how to work with `Graphs` and visualize them using Julia. In this section, we will briefly mention a few analysis tools that are available in the `Graphs` package and discuss some visualization opportunities.

`Graphs` offer functions for assessing *graph connectivity*. Among those functions, we can find the `connected_components` function for undirected graphs and the `strongly_connected_components` and `weakly_connected_components` functions for directed graphs. These functions return a vector of vectors containing the indices of the vertices that belong to a given component. If you want to visualize one of the connected components, you can use the `induced_subgraph` function to get the induced subgraph by the vertices in the corresponding component.

The `Graphs` package also offers an `articulation` function, which returns a vector with all the *cut vertices* in a connected graph. As we have a list of vertex indices as output, it is easy to highlight those vertices in the visualization using the vertex's color, size, shape, or label. The `bridge` function returns a vector of **bridges**, where edges that are deleted increases the number of connected components. Therefore, we can select those edges to highlight them in `GraphRecipes` by using the `src` and `dst` function to create the edge key tuple.

The `Graphs` package also offers a series of *shortest-path algorithms*, such as Dijkstra's algorithm, which is implemented in the `dijkstra_shortest_paths` function. These functions return Julia objects that store multiple pieces of information about how the graph traversed their fields. The *coloring algorithms* also return a special object of the `Coloring` type. These objects have a `num_colors` field for storing the number of colors and a `colors` field containing a vector that assigns a color, or an integer, to each vertex. When using `GraphRecipes`, you can pass the vector in the `colors` field to the `nodecolor` keyword argument if the number of colors in `num_colors` is equal to or less than the number of colors in the color palette.

`Graphs` also has functions for *measuring centrality*, such as `betweenness_centrality`. Most of these functions return a vector of floating-point numbers since there's a centrality value for each node. If we want to color the nodes according to their centrality using `GraphRecipes`, we need to pass that vector to the `marker_z` keyword argument rather than to `nodecolor`. If you want to see the color bar so that you have a reference, you also need to set `colorbar` to `true`.

In this section, we briefly introduced some of the analysis tools that the `Graphs` package offers. However, we can perform more operations on graphs using this package. For example, we have not talked about *distance measurements* and algorithms for detecting *community structures*. What's more, other packages from the *JuliaGraphs ecosystem* provide additional graph analysis tools.

Summary

In this chapter, we learned how to work with graphs in Julia using the `Graphs` package. We also explored the `MetaGraphs` package for storing graph, edge, and vertex metadata. Then, we learned how to visualize those graphs using the Julia ecosystem's three main packages: `GraphPlot`, `GraphRecipes`, and `GraphMakie`. The three packages, while similar, have different strengths. We learned about the main attributes available to customize their produced visualizations, and we saw examples of their strong points. Then, we saw how to change the graph layout of our visualization, and we briefly mentioned the analysis tools available in the `Graphs` package. With the knowledge you've acquired in this chapter, you can start visualizing and analyzing your graphs and networks using the Julia language.

In the next chapter, we will briefly introduce the tools in the Julia ecosystem for visualizing geographically distributed data.

Further reading

Since the `Graphs` package is extensive, we couldn't include all its tools in this chapter. We recommend looking into its documentation to explore the other methods it provides for working and analyzing graphs:

- You can find the `Graphs` documentation at `https://juliagraphs.org/Graphs.jl/stable/`.

- We couldn't expand more on all the interactivity offered by `GraphMakie`, but you can read more about it in its documentation: `http://juliaplots.org/GraphMakie.jl/stable/`.

- Finally, we recommend reading the `NetworkLayout` documentation when trying to customize the different layout algorithms since we haven't described all the available settings: `https://juliagraphs.org/NetworkLayout.jl/stable/`.

8

Visualizing Geographically Distributed Data

Visualizing geographically distributed data is vital for many scientific fields, such as ecology and climate sciences. It is essential to highlight patterns that arise from the spatial distribution of samples. Also, it is crucial to understand processes affecting the globe, such as climate change.

This chapter will briefly introduce the main `Plots` and *Makie* ecosystem packages for visualizing such data. In particular, we will learn how to plot choropleth and street maps.

In this chapter, we're going to cover the following main topics:

- Creating choropleth maps
- Introducing OpenStreetMapX

Technical requirements

For this chapter, you will need Julia, Pluto, and a web browser with an internet connection. In the `Chapter08` folder of the book's GitHub repository, you will find the Pluto notebooks with the code examples and their HTML versions with the outputs: `https://github.com/PacktPublishing/Interactive-Visualization-and-Plotting-with-Julia`.

In the `Chapter08` folder, you will also find the files containing the data we will visualize in this chapter: `snow.zip`, `countries.geojson`, and `map.osm`.

Creating choropleth maps

Choropleth maps are pretty standard in diverse types of media. Usually, they display established geographical entities, such as countries and provinces, using polygons. Following this, we can indicate the value of a variable for a region using the color inside its polygon. Usually, we will use the color or hue to identify the values of a categorical variable. For qualitative and ordinal variables, we use color luminance and saturation instead.

We can read the polygons that are needed to create choropleth maps from a vector file in one **Geographic Information System (GIS)** file format. The Julia language ecosystem offers the pure Julia `Shapefile` and `GeoJSON` packages to read files in the homonymous GIS formats. To read other GIS formats, both vector and raster images, you can use the `GDAL` or `ArchGDAL` packages. These last packages are wrappers to the **Geospatial Data Abstraction Library (GDAL)**, which can read and write multiple GIS formats.

The `Shapefile` and `GeoJSON` packages return a structure that follows the interface defined in the `GeoInterface` package. This last package also defines a series of Plots recipes for the geometrical entities defined by those formats. Therefore, you can call the `plot` function from `Plots` on the objects returned by those packages. In the *Makie* ecosystem, the `GeoMakie` package offers recipes to visualize geographically distributed data. It can plot out-of-the-box objects that the functions from the `GeoJSON` package return. As of `GeoMakie` version 0.3, you can also use shapefiles, but dealing with them is more cumbersome.

While `Shapefile` offers Plots recipes, you can have additional functionality through three main functions of the `PlotShapefiles` package:

- `plotshape`: This function allows you to plot the points, lines, and polygons defined in the shapefiles.

- `choropleth`: This function takes the polygons, a variable to determine the shading, and a color map to create a choropleth map.

- `google_overlay`: This function allows you to plot the previous maps over a static image from Google Maps.

The PlotShapefiles package uses Compose to create the visualizations and the Shapefile package to parse the input files.

There are two leading organizations in the Julia ecosystem for working with geographic data: *JuliaGeo* and *JuliaEarth*. The Shapefile and GeoJSON packages, along with the GeoInterface package, belong to the *JuliaGeo* organization. *Shapefiles* are composed of multiple files; some are optional. Still, they always have a file with the shp extension defining the geometries. Usually, they also have a dBase database, using the dbf extension, with information associated with the geometries in the shp file. The Shapefile package uses the DBFTables package, from the *JuliaGeo* organization, under the hood to allow the Table constructor to read such tables. In that organization, you will find some packages that we will discuss later in this chapter, such as the Proj4 package for cartographic projections.

Using some of the packages we mentioned earlier, we can create choropleth maps using Plots and *Makie*; we will see some examples later in this section. Other Julia packages, such as VegaLite and GMT, can also create choropleth maps, but we will not discuss them in this chapter. Let's see how to work with shapefiles and generate a choropleth map with Plots.

Creating a choropleth map with Plots

To explore how to work with Shapefile and create choropleth maps with Plots, we will use the data about the cholera outbreak in Soho, London, in 1854. That is a well-known dataset highlighting the importance of visualizing geographically distributed data. At that time, John Snow, an English physician, used that map to support his hypothesis that cholera is a water-borne disease. The spatial distribution of cases shown in that map allowed him to identify the Broad Street pump at the outbreak's center. The map used by John Snow was a *dot distribution map*, where he indicated the geographical location of the deaths caused by cholera using dots. Instead, we will create a *choropleth map* by adding the cholera-related deaths per block. Let's open a new Pluto notebook to start making that plot:

1. Execute the following command in the first cell:

    ```
    download("https://geodacenter.github.io/data-and-lab/
    data/snow.zip", "snow.zip")
    ```

    We have downloaded the data from GeoDa's sample datasets. In particular, we are downloading a ZIP file containing the data from the *Snow dataset*. If you get an error with that link, there is a copy of the dataset in the Chapter08 folder of the GitHub repository for this book; you can copy and paste the ZIP file into your current working directory and continue with the next step.

2. In a new cell, execute the following code:

```
using InfoZIP
```

Here, we loaded the `InfoZIP` package to decompress the ZIP files using the `unzip` function.

3. Execute the following code in a new cell:

```
InfoZIP.unzip("snow.zip")
```

We decompressed the ZIP file into the current working directory. This command creates a `snow` folder containing the dataset. Let's explore dataset number 3, with cholera deaths aggregated to blocks.

4. Run the following code line in a new cell:

```
readdir("snow/snow3")
```

We can see that the `snow` folder has a `snow3` folder containing dataset number 3. The folder includes the dataset in two formats, *GeoJSON* and *shapefile*. The GeoJSON file is `deaths_by_block.geojson`, while all the other files with `deaths_by_block` and different extensions are part of the shapefiles. The files with the `shp`, `dbf`, and `shx` extensions are mandatory for the *shapefile format*. The first file contains the geometric shapes, the second is the database storing the attributes for each shape, and the third is a positional index to navigate the first file quickly. In this case, the *Snow* dataset has a text file, with the `prj` extension, storing information about the used projection. Let's read some of these files using the `Shapefile` package.

5. In a new cell, execute `using Shapefile` to read the shapefiles.

6. Execute the following code in a new cell:

```
shp_block = Shapefile.Handle("snow/snow3/deaths_by_block.
shp")
```

We used the `Handle` constructor to read the shapefile with the `shp` extension into an object containing the shapes. In this case, we can see that the `shapes` field contains a vector of `Polygon` objects. Let's plot this file using the Plots recipe defined on the `GeoInterface` package.

7. In a new cell, execute the following code:

```
using Plots
```

We loaded the `Plots` package to access the Plots recipes.

8. In a new cell, execute the following code:

```
plot(shp_block)
```

When calling the plot function from Plots in the Handle object of Shapefile, Plots uses the Plots recipe defined in GeoInterfaces to plot the geometrical shapes in the shapefile. Here, we got a plot showing the different polygons representing the affected blocks in Soho, London. In this plot, the colors have no meaning, as Plots assigns a different discrete color from its palette to each polygon. The Plots package cycles those colors because the number of shapes is greater than the number of discrete colors in the palette, creating different blocks with the same color. The following figure shows the plot generated by this command:

Figure 8.1 – Shapefile plotted with Plots using the default attributes

9. In a new cell, execute the following code:

```
dbf_block = Shapefile.Table("snow/snow3/deaths_by_
block");
```

We use the Table constructor from Shapefile, which uses DBFTables under the hood to parse the file with the dbf extension. Note that we haven't written the dbf extension; we need to pass the path to the shapefile, along with the filename, optionally with the `shp` extension. The code returns a `Table` object that follows the table interface defined in the `Tables` package. We used the *semicolon* at the end of the line to suppress showing the output as Pluto cannot display that object. Let's create a `DataFrame` package with the information contained in the `Table` object.

10. Execute the following in a new cell:

```
using DataFrames
```

We need to load the `DataFrames` package to convert the `Table` object from `Shapefile` into `DataFrame`.

11. In a new cell, execute the following line of code:

```
data_block = DataFrame(dbf_block)
```

We created a `DataFrame` package from the `Shapefile` package. In the first column, named `geometry`, we have the shapes from the shapefile. In this case, the column has a collection of the `Polygon` objects defining the blocks. Each shape corresponds to a DataFrame's row.

12. In a new cell, execute the following code:

```
plot(data_block.geometry)
```

The `plot` function will also use the Plots recipe defined in `GeoInterface` when called using a vector of geometries. In this case, we plotted the `DataFrame` column with the `Polygon` objects to visualize them. As you can see, the generated plot is the same as the one shown in *Figure 8.1*. The created figure is not a choropleth map, as the colors don't have a meaning; let's make one.

13. In a new cell, execute the following code:

```
plt = plot(data_block.geometry,
    fill_z = permutedims(data_block.deaths),
    colorbar_title = "cholera deaths",
    seriescolor = :Greys_3,
    linecolor = :darkgray,
    framestyle = :none)
```

We used the same Plots recipe as shown in the previous step, but now we have also used attributes to customize our plot to create a choropleth map. The key attribute for achieving this is `fill_z`, as it assigns the fill color of the shapes according to the values in one variable. In this case, we color the blocks accordingly with the number of cholera deaths indicated by the `deaths` column of the `DataFrame` package. Each polygon counts as a series; therefore, we need a *row matrix* where each column passes a value for each polygon. To get the row matrix from the column vector, we use the `permutedims` function. Then, we use the `colorbar_title` attribute to document the meaning of the color scale. The other attributes are optional for our choropleth map; we have changed the default color palette to `:Greys_3` using the `seriescolor` attribute, `linecolor` to `:darkgray`, and hid the axis and box by setting `framestyle` to `:none`. The following figure shows the choropleth map we have generated:

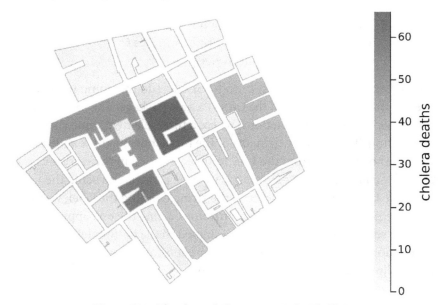

Figure 8.2 – The choropleth map created with Plots

Let's enhance this choropleth map by adding the locations of the pumps.

14. In a new cell, run the following code:

```
data_pumps = DataFrame(Shapefile.Table("snow/snow6/
pumps"))
```

We use the `Table` constructor to read the data in the shapefile from the `snow6` dataset. This shapefile contains the locations of the pumps. In the same command, we converted the table into a `DataFrame` package. This time, the `geometry` column has a set of `Point` objects.

15. Run the following code in a new cell:

```
plot!(plt, data_pumps.geometry, seriescolor=:black)
```

Here, we added the `:black` point, indicating the locations of the pumps as defined in the `geometry` column from the `DataFrame` package. Let's add the pump names to the plot.

16. In a new cell, run the following code:

```
annotate!(plt, [
    (row.geometry.x, row.geometry.y,
        text(row.name, 10, :bottom, :left))
    for row in eachrow(data_pumps) ])
```

We used the `annotate!` function from `Plots` to add the pump names, as indicated in the `name` column of the `DataFrame` package. The `annotate!` function takes a vector of annotations, where a three-element tuple characterizes each annotation. The two first elements of the tuple indicate the *x* and *y* coordinates. We got those coordinates from the `Point` objects stored in the `geometry` column. The third element is a `PlotText` object; we have created one using the `text` function. The first argument to `text` is the string defining the label; we used the strings in the `name` column. The other attributes define the text appearance; in this case, we set the font size to 10 points aligned to the bottom left. We created the vector of annotations through an array comprehension iterating the DataFrame's rows. The following figure shows the final plot:

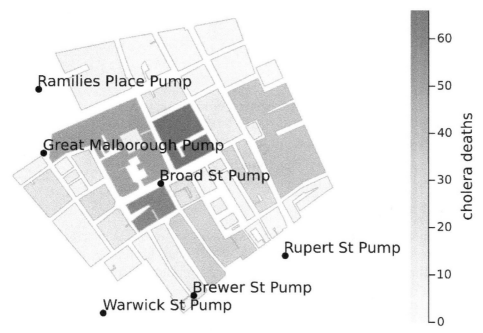

Figure 8.3 – The choropleth map of John Snow's dataset showing the locations of the pumps

We can now see the Broad Street pump at the center of the blocks showing the highest number of cholera deaths.

Great! We have created a choropleth map using `Plots` and `Shapefile`. Now, we will learn how to create one using `GeoMakie` and `GeoJSON`.

Creating a choropleth map with GeoMakie

In the *Makie* ecosystem, the `GeoMakie` package offers a series of functions and recipes to visualize geospatial data. Let's create a choropleth map from a *GeoJSON* file using `GeoMakie` in a new Pluto notebook:

1. In the first cell, execute the following code:

```
begin
    import Pkg
    Pkg.activate(temp=true)
    Pkg.add([
        Pkg.PackageSpec(name="GeoMakie", version="0.4"),
        Pkg.PackageSpec(name="CairoMakie",
version="0.8"),
```

```
        Pkg.PackageSpec(name="GeoJSON", version="0.5"),
    ])
    using GeoMakie, CairoMakie, GeoJSON
end
```

We have loaded `GeoMakie` to create our choropleth map using the `CairoMakie` backend. Additionally, we have loaded the `GeoJSON` package to parse files in the homonymous format. Note that this first cell can take some time to finish running.

2. In a new cell, execute the following command:

```
download("https://datahub.io/core/geo-countries/r/
countries.geojson", "countries.geojson")
```

This downloads the `countries.geojson` file containing the geographic boundaries for all the nations. We downloaded the data from the *DataHub* website. The data comes from the *Natural Earth* website, which is a database of public domain maps. There is a copy of the `countries.geojson` file in the `Chapter08` folder of this book's GitHub repository. If you find an error with this command, copy the file from there and paste it into the current working directory.

3. In a new cell, execute the following code:

```
countries = GeoJSON.read(read("countries.geojson"))
```

We have parsed the information inside the GeoJSON file, creating a `FeatureCollection` object from the `GeoInterface` package. If you explore that object in Pluto, you will notice that it contains a vector of `Feature` objects in the `features` field. Each `Feature` has a `geometry` field and a `properties` field; the former has the geometrical shapes, and the latter contains a `Dict` with metadata. In the following steps, we will use that data to color our choropleth map.

4. Execute the following code in a new cell:

```
name_lengths = [
    length(feature.properties["ADMIN"])
    for feature in countries.features]
```

We used array comprehension to create a vector containing the number of characters in the English name of each country. We iterated each `Feature` in the `features` vector, and we took the `ADMIN` value from the `properties` dictionary.

5. Run the following code in a new cell:

```
poly(countries, color=name_lengths)
```

The easiest way to create a choropleth map with *Makie* is to call the `poly` function using the object returned by `GeoJSON`. We used the `color` attribute to indicate the variable we wanted to use to color the shapes. In this example, *Makie* maps a country's name length to a color on the default color gradient, `:viridis`. The following figure shows the resulting plot:

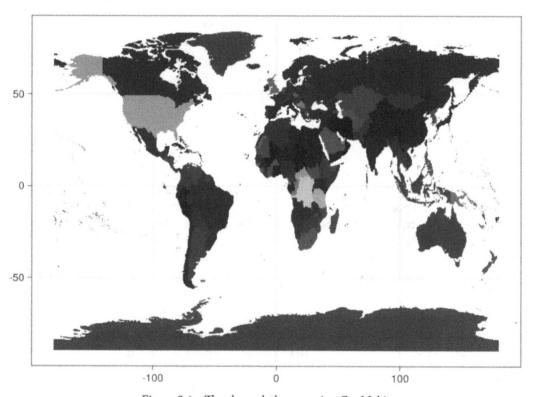

Figure 8.4 – The choropleth map using GeoMakie

6. In a new cell, execute the following code:

```
begin
    fig = Figure()
    ax = GeoAxis(fig[1,1],
        coastlines = true,
        coastline_attributes = (color=:black,),
        dest = "+proj=natearth")
```

```
    poly!(ax, countries, color= name_lengths)
cb = Colorbar(fig[1,2];
        colorrange = extrema(name_lengths),
        label = "country name length",
        height = Relative(0.65))
    fig
end
```

We have taken advantage of `GeoAxis`, which `GeoMakie` defines, for plotting geospatial data. First, we created an empty figure; we will lay out our map and its color bar into that figure. Then, we made our `GeoAxis`, and we located it in the first row and column of the figure layout. All the keyword arguments for `GeoAxis` are optional. Here, we have chosen to show `coastlines` using the `:black` color. The `coastline_attributes` keyword argument passes arguments to the `lines` function, which draws the coastlines using a named tuple; that's why we need the trailing comma after the color name.

One of the prominent aspects of `GeoAxis` is that it allows you to change the *map projections*. We can use three keyword arguments to change the map projection: `source`, `dest`, and `transformation`. The three keywords use the `Proj4` package, a wrapper to the *PROJ* library, to perform the conversions. The `source` and `dest` keyword arguments, respectively, define the projection used in the input data and the one desired for the output. In the last step of the example, we set the `dest` keyword argument to `"+proj=natearth"` to use a *Natural Earth* projection instead of the default *Equal Earth* projection. You can find the list of map projections and the syntax to define them in the *PROJ documentation*.

In *step 6*, after creating `GeoAxis`, we have used the `poly!` function to plot the `Polygon` objects from the `FeatureCollection` instance, as created in *step 3*, into the axes. We used the `color` attribute of `poly!` to color the polygons according to the country name length. Finally, we added a color bar to our choropleth map in the second column of our figure using the `Colorbar` constructor. We defined the range of colors using the `colorrange` attribute to match the actual extreme values in our variable. The following figure shows the created plot:

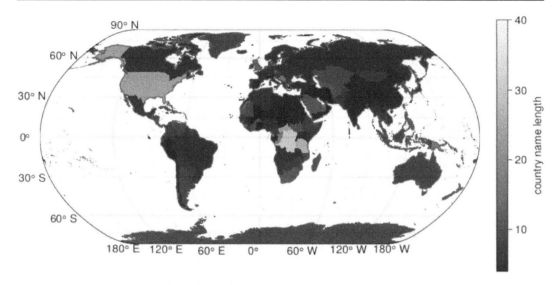

Figure 8.5 – The choropleth map using GeoAxis

In the previous example, we only used the `poly!` function from *Makie* to draw the polygonal shapes. However, as we do for `Axis`, we can call any other *Makie* function on `GeoAxis`. For example, we can plot points in the map using the `scatter!` function. When using the `CairoMakie` backend, we can use the `image!` and `heatmap!` functions. For example, you can execute the following code in a new cell at the end of the Pluto notebook we used in this section:

```
begin
    earth_fig = Figure()
    earth_ax = GeoAxis(earth_fig[1,1],
        lonlims = automatic,
        dest = "+proj=ortho")
    image!(earth_ax, -180..180, -90..90,
        rotr90(GeoMakie.earth());
        interpolate = false)
    earth_fig
end
```

Here, we used the `image!` function to plot a picture of Earth obtained from the `earth` function of `GeoMakie`. In this case, we plotted the image using an orthographic projection — thanks to `proj=ortho` in the `PROJ` string we passed to `dest`. Note that we have set the `lonlims` attribute to `automatic` so that *Makie* can fit the best longitude limits for the *x* axis. The `image!` function takes, optionally, an object where to draw our image. In this case, we plotted the Earth image into the `GeoAxis` we just created. Then, `image!` takes the longitude and latitude values using positional arguments. After those, we pass the data we want to plot on the map. For this kind of plot, we need to set the `interpolate` attribute to `false` — otherwise, we will see an error. Note that we needed to rotate the Earth image 90 degrees using the `rotr90` function. That code block created the image shown in the following figure:

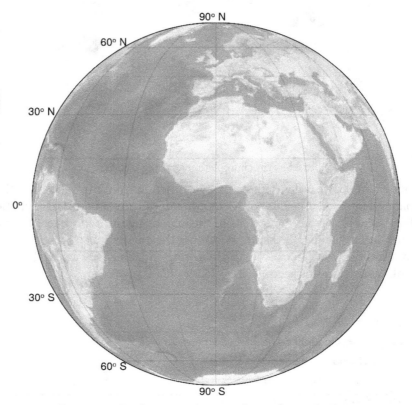

Figure 8.6 - Earth image using an orthographic projection.

Note that if we use the `GLMakie` backend, we should use the `surface!` function instead of the `image!` and `heatmap!` functions.

In this section, we have learned the basics of plotting geographically distributed data with `Plots` and *Makie* using files in the shapefile and GeoJSON formats. In particular, we have seen how to create choropleth maps. In the next section, we will see how to access and plot information from OpenStreetMap.

Introducing OpenStreetMapX

OpenStreetMap is a collaborative project in which to create and keep updating a digital map of the world that you can explore at www.openstreetmap.org. You can download the map of a desired region of the world in the *OSM format*—an XML file—from that site. In Julia, you can parse OSM files using the OpenStreetMapX package. Then, you can visualize it using Plots, the default option, PyPlot, through the OpenStreetMapXPlot package, or using the folium Python package, through PyCall. Let's quickly explore the OpenStreetMapX and OpenStreetMapXPlot packages by plotting a map of present-day Soho, London, using Pluto:

1. In the first cell, execute the following code:

    ```
    using OpenStreetMapX, OpenStreetMapXPlot, Plots
    ```

    This loads the necessary library to plot the data from OpenStreetMap in Julia using Plots.

2. Execute the following code in a new cell:

    ```
    Download("https: https://github.com/PacktPublishing/
    Interactive-Visualization-and-Plotting-with-Julia/raw/
    main/Chapter08/map.osm", "map.osm")
    ```

    This code downloads the map.osm file from the Chapter08 folder of the GitHub repository of this book into the current working directory.

3. In a new cell, run the following code:

    ```
    soho = get_map_data("map.osm", use_cache = false)
    ```

    The get_map_data function from OpenStreetMapX parses the data in the OSM file and returns a MapData object. Here, we set use_cache to false to avoid storing a local copy of the requested data.

4. In a new cell, run the following line of code:

    ```
    soho.bounds
    ```

    Here, we can see that the bounds field of the MapData object has the bounding box for our map in **Latitude-Longitude-Altitude (LLA)** coordinates.

5. In a new cell, execute the following code line:

    ```
    p = plotmap(soho)
    ```

The `plotmap` function from `OpenStreetMapX` can take the `MapData` object and plot the stored map. The map uses the **East, North, Up (ENU)** coordinates, with the box's center as *(0, 0)*. By default, the distances are in *meters (m)*, but you can set the km keyword argument to `true` to use kilometers instead. Let's add a point where the Broad Street pump is located.

6.   In a new cell, run the following code:

```
pump = ENU(LLA(51.51334705377814, -0.1366063043462497),
OpenStreetMapX.center(soho.bounds))
```

We used the `LLA` constructor from `OpenStreetMapX` to create a point using the LLA coordinates. Then, we used OpenStreetMapX's `ENU` constructor to convert that point from LLA coordinates into ENU ones. Note that we need to give the LLA coordinates of the reference point, the center of the bounding box. Therefore, we applied the `center` function from `OpenStreetMapX` to the `bounds` field of the `MapData` object to calculate the location of the reference point. Finally, we get the location of the Broad Street pump in ENU coordinates.

7.   In a new cell, execute the following code:

```
scatter!(p, [pump.east], [pump.north])
```

We added a point using the `scatter!` function at the location of the Broad Street pump. We need to give vectors of *x* and *y* coordinates that correspond to *east* and *north* in the case of ENU coordinates. We got those from the `east` and `north` fields of the `ENU` object. The following figure shows the resulting plot, which is from OpenStreetMap contributors:

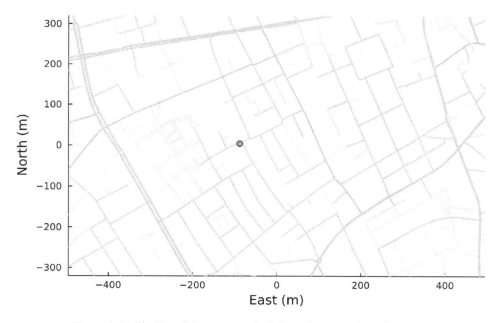

Figure 8.7 – The Broad Street pump in Soho using OpenStreetMapXPlot

In this section, we have seen how to plot a map obtained from OpenStreetMap and highlight a location from another coordinate system. This knowledge will be advantageous when visualizing up-to-date maps, particularly at the city level.

Summary

In this chapter, we learned how to create choropleth and street maps in Julia. We can now read maps in the shapefile, GeoJSON, and OSM formats. Additionally, we saw the basic requirements to plot geographically distributed data using `Plots`, `GeoMakie`, and `OpenStreetMapXPlot`. The three packages have different strengths and use cases, for which we have explored the most classic ones. Therefore, this chapter should be a good starting point when trying to plot our own geographically distributed data.

In the next chapter, we will take advantage of the available packages in the Julia ecosystem for plotting biological data.

Further reading

When willing to change map projections with the `GeoMakie` and `Proj4` packages, you will find that the *PROG* documentation is highly useful: `https://proj.org/`.

Also, as `GeoMakie` is a project in development, you can find it useful to refer to its documentation: `https://juliaplots.org/GeoMakie.jl/stable/`.

9
Plotting Biological Data

Technical advancements, such as new sequencing technologies or machine learning algorithms, are driving a revolution in biology. That revolution is expanding the fields of bioinformatics and computational biology. The large amount of biological data we are generating nowadays also pushes innovation on the data visualization front. Interactive visualizations are essential for navigating such extensive and complex data. Biological information is also highly heterogeneous; we can have sequences, medical images, and geographical locations, to name a few. Biological data visualization is a vast subject, from which we will only explore ways to visualize phylogenetic trees and protein sequences and structures in this chapter. However, the Julia ecosystem has more tools for analyzing and visualizing biological data. You can explore some of those tools in the *BioJulia*, *JuliaHealth*, and *EcoJulia* organizations.

In this chapter, we will learn how to visualize phylogenetic trees using Julia. We will also explore using the `MIToS` package to work with proteins. We will discuss different Julia packages for visualizing protein sequences and structures, including ways to examine multiple sequence alignments. Finally, we will learn how to create interactive dashboards for biological data using `BioDash`.

In this chapter, we're going to cover the following main topics:

- Visualizing phylogenetic trees
- Plotting a protein sequence and structure
- Creating dashboards for biological data

Technical requirements

For this chapter, you will need the following:

- Julia 1.6 or higher with the `Pluto` package installed
- A web browser and an internet connection
- A text editor could be handy to work with `BioDash`; we recommend having Visual Studio Code with the Julia extension installed

The code examples and the Pluto notebooks, including their HTML versions with embedded outputs, for this chapter can be found in the `Chapter09` folder of this book's GitHub repository: `https://github.com/PacktPublishing/Interactive-Visualization-and-Plotting-with-Julia`.

Visualizing phylogenetic trees

A **phylogeny** or **phylogenetic tree** aims to show the evolutionary relationships between biological entities, such as species or genes. Trees, in graph theory, are fully connected undirected graphs without cycles; therefore, we can use software aimed at graph visualization to explore phylogenies. In fact, in *Chapter 7, Visualizing Graphs*, we used `GraphRecipes` to visualize other trees, namely, abstract syntax trees and type hierarchies. However, if we want to visualize phylogenies, we should rely on dedicated packages. The main package for working with phylogenies in the Julia ecosystem is `Phylo`, from the EcoJulia organization. This package offers a `Plots` recipe for their tree objects. In this section, we will explore the `Phylo` package and phylogenetic trees. However, if you deal with phylogenetic networks, you should rely on the `PhyloNetworks` and `PhyloPlots` packages. **Phylogenetic networks** are more complex graphs than trees in that a node can have more than one parent. Therefore, they allow you to express events such as lateral gene transfers or recombinations.

We can store and share phylogenies using the Nexus and Newick file formats. The `Phylo` packages can parse both, utilizing the `parsenexus` and `parsenewick` functions. We can input this function with a `String` containing the tree in the corresponding format or as the first positional argument for the `open` function to parse a tree contained in a file.

The returned object is a subtype of `AbstractTree` that the `Plots` recipe that's exported by the package can plot. Therefore, we can call the `plot` function in the returned object to visualize the tree.

If you read a Nexus file that stores multiple trees, the returned object will be of the `TreeSet` type; you will need to index each tree by its name to plot it. The `Plot` recipe has a series of extra keyword arguments that can come in handy when dealing with big trees. For example, you can change the default `treetype` from `:dendrogram` to `:fan`, suppress the leaf labels by setting `showtips` to `false`, or reduce their size using the `tipfont` attribute. The other two attributes are `marker_group` and `line_group`, which apply the `group` attribute to node markers and branch lines, respectively.

Let's plot a simple tree in Newick format using the `Phylo` package:

1. Open a new Pluto notebook and execute the following in the first cell:

    ```
    using Phylo, Plots
    ```

    Here, we loaded `Phylo` to work with the phylogeny and `Plots` to access the plot recipe.

2. In a new cell, execute the following code:

    ```
    newick = "(Leaf_A:0.1,(Leaf_B:0.2,Leaf_C:0.1)
    Internal:0.1)Root;"
    ```

    The preceding code creates a `String` object containing a tree in Newick format.

3. In a new cell, execute the following code:

    ```
    tree = parsenewick(newick)
    ```

    Here, we used the `parsenewick` function to parse the tree from the previously created `String`. The function returned a `RootedTree` object.

4. Execute the following code in a new cell:

    ```
    plot(tree, size=(500,200))
    ```

    Here, we used the `plot` function to visualize `RootedTree`. As the recipe creates a plot that's `(1000, 1000)` in size by default, we used the `size` attribute to create a smaller tree. The tree is similar to the one shown to the left of *Figure 9.1*.

5. In a new cell, execute the following code:

    ```
    begin
        before = plot(tree,
            title="Tree before sorting",
    ```

```
                    size=(500,200))
        sort!(tree)
        after = plot(tree,
            title="Tree after sorting",
            size=(500,200))
        plot(before, after,
            titlefontsize=8, titlelocation=:left)
    end
```

First, we created a tree plot like the previous one, but with a title. Then, we *ladderized* our tree using the sort! function as it sorts the tree branches at each node according to the number of descendants. After that, we plotted the tree again to see the new order. Finally, we created a panel containing both plots to compare the differences. The following diagram shows the created plot; note that the order of the leaves on the left-hand side can change from run to run:

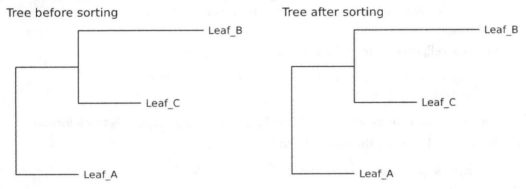

Figure 9.1 – Phylogenetic tree before and after branch sorting

6. In a new cell, execute the following code block:

```
trait = map_depthfirst(
    (val, node) -> val + randn(),
    0.0, tree)
```

Here, we created a trait vector representing a continuous genetically determined character using the map_depthfirst higher-order function. The map_depthfirst function applies the function we pass in the first positional argument to each tree node in depth-first order. The input function should take two arguments; the value returned from the function on the previous node and the current node. In this case, we defined an anonymous function that adds a random number to the value of the last node. For the first node, we set the starting value to `0.0` using the second positional argument. The third and final positional argument of the `map_depthfirst` function is the tree to be iterated.

7. In a new cell, execute the following code:

```
plot(tree,
    treetype = :fan,
    line_z = trait,
    linewidth = 3,
    linecolor = :RdYlGn_4,
    size = (450,200))
```

Here, we plotted the tree using a *circular fan-like layout* by setting the `treetype` attribute to `:fan`. We colored the branches by setting the `line_z` attribute to the trait vector defined in the previous step. Using the `linewidth` attribute, we made the lines thicker to make the colors easier to see. Finally, we changed the color scheme for the branches using the `linecolor` attribute. The following diagram shows the plot:

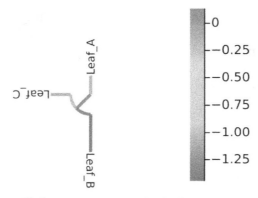

Figure 9.2 – Phylogenetic tree using the fan layout and branch colors

In this section, we learnt the basics of plotting phylogenetic trees with Julia using the `Phylo` package. In the next section, we will learn how to explore protein sequences and structures visually.

Plotting a protein sequence and structure

Proteins are the leading actors in the cell, performing most of the actions needed to sustain life. These macromolecules are amino acid residue chains. We usually represent amino acid residues using characters and, therefore, a protein chain as a sequence of characters. Protein chains can adopt a set of tridimensional structures that we can analyze to understand their functional mechanism. This section will explore ways to visualize protein sequence and structure using MIToS and packages from the BioJulia organization. In particular, we will not focus on exploring single sequences but on **multiple sequence alignments (MSAs)**. These alignments show information about the sequence of interest and many other evolutionary-related sequences.

While MIToS focuses on analyzing proteins, the packages in the BioJulia ecosystem can also model other macromolecules, especially DNA and RNA molecules. Using MIToS, we can take advantage of its Plots recipes to quickly visualize its objects. When using MIToS or the BioJulia packages, we can take advantage of the BioMakie package, which offers Makie recipes for visualizing protein structures and multiple sequence alignments. Also, when visualizing structures from MIToS or BioStructures, you can use the Bio3DView package, a Julia wrapper for the 3Dmol.js JavaScript library. Both BioMakie and Bio3DView produce interactive plots. The DashBio package also offers interactive utilities for visualizing sequences, alignments, and structures, as we will see in the next section.

Exploring multiple sequence alignments

We can load an MSA into Julia using the MIToS package, creating an alignment object and, by default, storing its metadata. The MIToS package offers a Plots recipe to observe the gap and conservation patterns in the alignment quickly. For a closer and more interactive look at the alignment, you can rely on the BioMakie package. Let's explore both options using a Pluto notebook:

1. In the first cell of the new notebook, execute the following code:

```
begin
    import Pkg
    Pkg.activate(mktempdir())
    Pkg.add([
        Pkg.PackageSpec(name="Plots", version="1"),
        Pkg.PackageSpec(name="MIToS", version="2"),
        Pkg.PackageSpec(name="BioMakie",
            version="0.2.0"),
    ])
    using Plots, MIToS.Pfam, MIToS.MSA, BioMakie
end
```

This cell loads a couple of packages with many dependencies; therefore, this cell could take minutes to load, especially the first time.

We have used Pluto's **Pkg cell** pattern to install specific versions of the desired packages. The pattern uses the Pkg module, mainly the add function and the PackageSpec type, to set up the temporal environment that's created with the activate function. You can read more about it in the *Managing environments* section of *Chapter 1, An Introduction to Julia for Data Visualization and Analysis*.

In this case, we installed MIToS to load Pfam and the MSA module. We need the former to download a multiple sequence alignment from the Pfam database and the latter for working with that alignment. We also loaded Plots, to access the plotting recipe from MIToS and BioMakie, to render the alignment with Makie.

Important Note

We loaded BioMakie 0.2.0 as the code example will not work with the newer version — at the time of writing, BioMakie is under active development.

---

2. Execute the following code in a new cell:

```
msa_file = downloadpfam("PF00049")
```

Here, we used the downloadpfam function from MIToS.Pfam to download the multiple sequence alignment for the *Insulin* protein family from the Pfam database using its accession, *PF00049*.

3. In a new cell, run the following code:

```
msa = read(msa_file, Stockholm)
```

Here, we read the multiple sequence alignment from the file that was downloaded from the Pfam database. As the file uses the *Stockholm format*, we used the second positional argument of the read function to indicate that. The read method from MIToS returns an AnnotatedMultipleSequenceAlignment object that we can plot.

4. In a new cell, run the following code block:

```
plotly()
```

Here, we set the Plots backend to Plotly to allow us to zoom into and pan the MSA plot. When loading the Plotly backend, Plots will tell us that we will need to load PlotlyBase if we want to export the Plotly figures as PNG files. We can ignore that message in this example, as we do not plan to save the plots.

5. Run `plot(msa)` in a new cell to see the multiple sequence alignment. It uses the `plot` function from `Plots` to access the plotting recipe defined in the `MSA` module of `MIToS`. The following diagram shows the created plot:

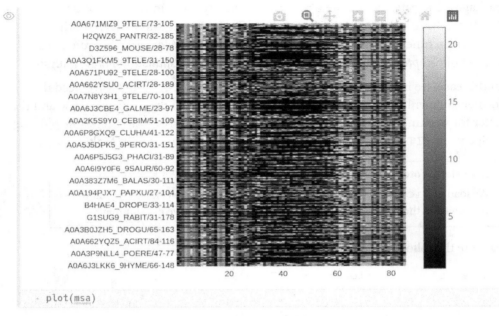

```
plot(msa)
```

Figure 9.3 – MSA using the MIToS Plots recipe

Here, we can see that the gaps, representing residues that have been inserted or deleted, are shown in black. Columns with the same color across different rows allow you to identify the conserved columns in MSA. While we can take advantage of the Plotly backend to zoom into the plot and move across it, as of `MIToS` 2.9, we cannot see the identity of the amino acid residues. Let's use `BioMakie` for that.

6. Run the following code in a new cell:

```
seqs = [ (name, stringsequence(msa, name))
        for name in sequencenames(msa) ]
```

Since `BioMakie` 0.2.0 does not automatically plot alignment objects from `MIToS`, we used array comprehension to create the necessary input. We used the `sequencenames` function to get a vector with the sequence names. Then, we iterated through it to create a new vector containing a tuple of `String` objects. Each tuple has the sequence's name in the first position and the aligned sequence as a `String` in the second. We used the `stringsequence` function to get the string containing the aligned sequence, given an MSA and a sequence name.

7. In a new cell, run the following line of code:

```
fig = viewmsa(seqs)
```

Here, we used the viewmsa function to create our figure. This displays a static figure of the MSA's first rows, sequences, and columns. BioMakie creates the figure using GLMakie, so let's open an interactive GLFW window.

8. Execute the following code in a new cell:

```
display(fig)
```

Here, we have opened an interactive window where we can use the row and column sliders to navigate the MSA. The following screenshot shows the visualization that was created:

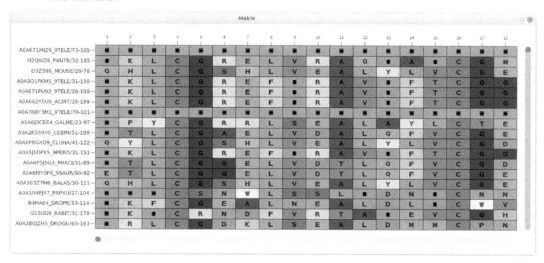

Figure 9.4 – MSA using MIToS and BioMakie

Now that we have learned how to visualize MSAs, let's learn how to visualize protein structures.

Visualizing protein structures

We can find experimentally determined protein structures in the **Protein Data Bank** (**PDB**). In this database, we can also find DNA and RNA structures. The main packages in the Julia ecosystem for working with structures from that database are MIToS, particularly its PDB module, and BioStructures from the BioJulia organization. The aforementioned BioMakie package offers a way to visualize the structures from BioStructures. This section will use the Plots recipe from MIToS.PDB to understand the global structure and Bio3DView for more detailed visualization. Let's create a new Pluto notebook to start exploring these packages:

1. Execute the following code in the first cell:

    ```
    begin
        using MIToS.PDB, Plots
        plotly()
    end
    ```

    Here, we loaded the PDB module from the MIToS package to work with the protein structures. We also loaded the Plots package to access the *Plots recipe*. Finally, we selected the *Plotly* backend to benefit from its interactivity.

2. In a new cell, execute the following code:

    ```
    pdbfile = downloadpdb("1SDB")
    ```

    Here, we downloaded the structure of the porcine insulin from the PDB database using the downloadpdb function. This function takes a String object containing the PDB identifier of the desired structure.

3. In a new cell, execute the following code:

    ```
    pdbres = read(pdbfile, PDBML)
    ```

    Here, we used the read function to parse the file downloaded from PDB into a vector of PDBResidue objects. The second positional argument indicates the file format. In this case, this is PDBML, the XML format from PDB that MIToS downloads by default.

4. Execute the following code in a new cell:

```
plot(pdbres)
```

Here, we used the `Plots` recipe defined in the `PDB` module of `MIToS` to trace the position of the alpha carbons in the protein to get a rough idea of the structure. Let's use `Bio3DView` to get a better visualization of it.

5. In a new cell, execute `using Bio3DView` to load that package.

6. Execute the following code in a new cell:

```
filter!(res -> res.id.name != "HOH", pdbres)
```

This step is not mandatory. Here, we discarded all the water molecules using the `filter!` function to get a nicer surface figure at the end. We have kept all the residues whose names are not `HOH`, as indicated in the `name` field of the `PDBResidueIdentifier` object, stored in the `id` field of each `PDBResidue`.

7. Execute the following code line in a new cell:

```
pdb_string = sprint(print, pdbres, PDBFile)
```

Here, we created a `String` object using the `sprint` function, which contains the PDB structure in the classic PDB format from the PDB. We need this string to plot it with `Bio3DView` as version 0.1.3 doesn't plot MIToS objects out-of-the-box. In this case, the `sprint` function captures the output of `print(pdbres, PDBFile)` in a `String`. The `PDBFile` type indicates that we want to use the PDB format.

8. In a new cell, run the following code line:

```
viewstring(pdb_string, "pdb")
```

Here, we used the `viewstring` function from the `Bio3DView` package to visualize the structure stored in our string. We used the second positional argument to indicate the format in which the string stores the structure, `pdb`. This function can take some optional keyword arguments, such as `axes` and `cameraangle`, to control the visualization. These keyword arguments usually take an object of a type with the same name but in camel case; for example, `Axes` and `CameraAngle`. We will see some of these keyword arguments in the following steps. The following figure shows the plot that was created in this step:

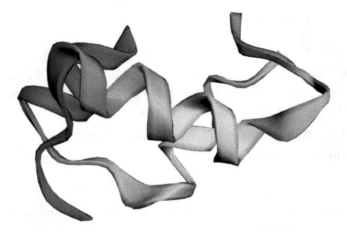

Figure 9.5 – PDB structure displayed by Bio3DView

9. In a new cell, execute the following code:

```
viewstring(pdb_string, "pdb",
    style=Style("stick",
        Dict("colorscheme"=> "chain")))
```

Here, we changed the visualization's `style` from the default of `cartoon` to `stick`. Other standard options are `line` and `sphere`. We indicated this with an object of the `Style` type. The type constructor can take an optional `Dict` of options that `Bio3DView` passes to the underlying JavaScript library. Therefore, you can look for details in the `3dmol.js` documentation if needed. Here, we used the dictionary to make the colors match the chains.

10. In a new cell, execute the following code block:

```
viewstring(pdb_string, "pdb",
    style=Style("stick",
        Dict("colorscheme"=> "chain")),
    surface=Bio3DView.Surface(
        Dict("opacity"=> 0.75, "colorscheme"=>
"chain")))
```

The preceding code is identical to the previous one, but we have also added a surface to our plot using the `surface` keyword argument. This argument takes an object of the `Surface` type. Since both `Plots` and `Bio3DView` export `Surface`, we needed to use the fully qualified name here; that is, `Bio3DView.Surface`. The `Surface` constructor can also take an optional dictionary with options; in this case, we used it to add transparency to the surface and make the surface follow the same colors as the sticks. The following figure shows the plot that was created:

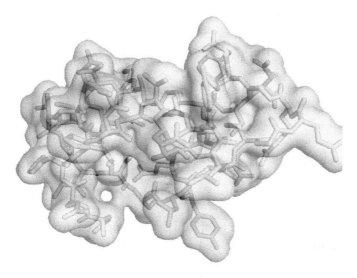

Figure 9.6 – Protein structure with a custom style and surface using Bio3DView

This section taught us how to visualize protein structures and multiple sequence alignments using MIToS, BioMakie, and Bio3DView. In the next section, we will learn how to create these visualizations and others using BioDash.

Creating dashboards for biological data

This section will teach us the basics of creating a Dash interactive dashboard to visualize biological data. We learned how to create a simple Dash application in the *Creating applications to serve interactive plots* section of *Chapter 3, Getting Interactive Plots with Julia*. For that, we used the Dash package. This section will expand on that to include the DashBio package. DashBio exports a series of components to visualize biological data. All the components start with the dashbio_ prefix; for example, the component to visualize multiple sequence alignments is dashbio_alignmentchart. The components are diverse; you will find tools to visualize sequences, chromosomes, molecular structures, and other bioinformatics-related plot types. Those components require different input data formats; we recommend checking the documentation to find the expected format.

As an example, we will create a Dash application that will display a protein structure using one of the many structure visualization components in `DashBio`. Follow these steps:

1. Open VS Code and create a new Julia file named `biodash_example.jl`. We will write the application in that file in the following steps.

2. At the beginning of the file, paste the following code block:

    ```
    import Pkg
    Pkg.activate(temp=true)
    Pkg.add([
        Pkg.PackageSpec(name="MIToS", version="2.9"),
        Pkg.PackageSpec(name="Dash", version="0.1"),
        Pkg.PackageSpec(name="DashBio", version="0.7"),
    ])
    ```

    The preceding code will create a temporal development environment for our application. In this environment, we have installed specific versions of the necessary packages to ensure the example will continue working in the future. In particular, we have installed `MIToS` to work with the protein structure and `Dash` and `DashBio` to create the dashboard.

3. Insert the following code block after the previous one:

    ```
    using MIToS.PDB
    using Dash, DashBio
    ```

    The preceding code loads the libraries we installed in the previous step.

4. Insert the following code block:

    ```
    pdbfile = downloadpdb("1SDB")
    pdbres = read(pdbfile, PDBML)
    ```

    Here, we used the `PDB` module of `MIToS` to download the *1SDB* structure from PDB using the `downloadpdb` function. Then, we parsed that file in `PDBML` format using the `read` function. We now have a `Vector` of `PDBResidue` objects. However, the `dashbio_speck` component requires the input structure to be a vector of dictionaries, one for each atom. Let's create those dictionaries now.

5. Add the following code block to the file:

    ```
    to_dict(atom) = Dict(
        "symbol" => atom.element,
        "x" => atom.coordinates.x,
    ```

```
    "y" => atom.coordinates.y,
    "z" => atom.coordinates.z)

data = collect(Iterators.flatten([
    [to_dict(atom) for atom in res.atoms]
    for res in pdbres]))
```

The preceding code block creates the necessary vector of dictionaries from the vector of PDBResidue objects. Each PDBResidue has an atoms field containing a vector of PDBAtom objects. Here, we are using **array comprehensions** to iterate over the residues and then iterate over the atoms for each residue. Then, we are applying the to_dict function we defined at the beginning of the code block to convert the PDBAtom object into the dictionary that dashbio_speck requires. This dictionary should have a "symbol" key to define the atom's chemical element and the "x", "y", and "z" keys to determine the atom's spatial coordinates. We take those values from the element and coordinates fields of the PDBAtom object. Note that the nested array comprehension will create a vector of vectors containing dictionaries rather than a vector of dictionaries. To solve this, we used the flatten iterator from the Iterators module to flatten the vectors, and then we collected the results.

6. Add the following code block at the end of the file:

```
app = dash()
app.layout = dashbio_speck(data=data)
run_server(app, "0.0.0.0", debug=true)
```

The preceding code block creates the dashboard with the interactive visual component. First, we used the dash function to generate the Dash application. Then, we add the desired component, dashbio_speck, to the application's layout. We input our data and let all the configurations be the default values. The components have a series of keyword arguments that we can use to configure the visualization. You can check the function documentation to learn more about those parameters. Finally, we used the run_serve function to run the app.

7. Press *Alt + Enter* to run the entire file. You will see the execution progress in the **TERMINAL** tab of Visual Studio Code.

8. Go to your web browser and enter the address that appears after **Listening on:** in the **Terminal** tab. It should be something like 0.0.0.0:8050. After that, you will see the application in the browser. The following screenshot shows the plot that was created; you can zoom out using the mouse wheel:

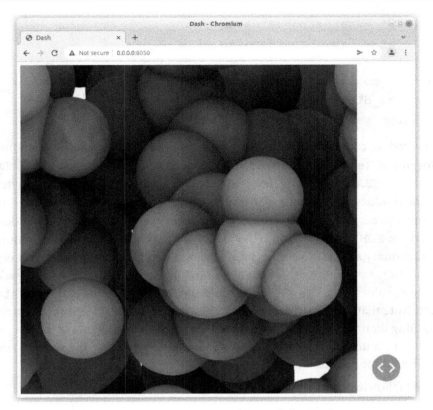

Figure 9.7 – Protein structure using DashBio and the Speck component

This example shows how to create and run a simple dashboard using interactive bioinformatics components thanks to the `DashBio` package. It should be helpful as a guide for creating similar applications to visualize other objects or share biological data observations.

Summary

This chapter taught us about some packages for working with and visualizing biological data. In particular, we learned how to plot phylogenetic trees, protein structures, and multiple sequence alignments. We learned how to explore the last two using different complexity levels by taking advantage of various packages from the Julia ecosystem, such as `MIToS`, through Plots recipes, or `BioMakie` in the Makie ecosystem. Lastly, we saw an example of how to use interactive components from the `DashBio` package, which offers a large spectrum of bioinformatics-related visualizations.

In the next chapter, we will learn about the different plot parts when working with `Plots`, `Makie`, and `Gadfly`. This will be extremely useful when customizing our figures.

Further reading

If you want to know more about how to use Julia in the field of biology, we recommend reading the article *Roesch, Elisabeth, et al. "Julia for biologists." arXiv preprint arXiv:2109.09973 (2021)*: `https://doi.org/10.48550/arXiv.2109.09973`.

Biological data visualization is a big topic, with many dedicated packages in the Julia ecosystem, and we have only explored a few. Even for these few packages, what we can do is more of what we have seen in this chapter. If you want to learn more about the different packages, we recommend looking into their documentation:

- You can find the `MIToS` documentation at `https://diegozea.github.io/MIToS.jl/latest/`.

- For customizing `Bio3Dview` plots, you can check out the documentation for the underlying JavaScript library, *3Dmol.js*: `https://3dmol.csb.pitt.edu/doc/index.html`.

- You can read more about the `DashBio` components at `https://dash.plotly.com/julia/dash-bio`.

- If you want to see the `DashBio` components in action, you can find them here: `https://dash.gallery/Portal/?search=[Biotechnology]`.

Section 3 – Mastering Plot Customization

In this section, we will learn how to customize Julia plots. These chapters will focus on the `Plots` package, but they will also hint at the essential customization aspects of `Gadfly` and *Makie*. First, we will learn about the different elements that compose a plot and the vocabulary the three packages use to refer to them. Knowing those elements is crucial, as later, we will learn about the different plot attributes available to customize them. Next, this section will teach us about the layout system of those packages to arrange multiple plots into a single figure. Then, we will customize essential plot elements and explore the use of color. After that, we will learn how to create and use plot themes – a valuable tool for reusing customizations. Finally, we will learn how to draw custom forms and create entirely new plot types thanks to the recipe system of `Plots` and *Makie*.

This section comprises the following chapters:

- *Chapter 10, The Anatomy of a Plot*
- *Chapter 11, Defining Plot Layouts to Create Figure Panels*
- *Chapter 12, Customizing Plot Attributes – Axes, Legends, and Colors*
- *Chapter 13, Designing Plot Themes*
- *Chapter 14, Designing Your Own Plots – Plot Recipes*

10
The Anatomy of a Plot

This is the first chapter of this book's third part, *Mastering Plot Customizations*. These last few chapters will allow you to fully customize Julia's plots – mainly via the `Plots` package. To customize our plots, we should first know the various parts that compose them. Those components can have different names and customization levels across plotting packages.

Therefore, this chapter will dissect figures from `Plots`, *Makie*, and `Gadfly` to understand their terminology. This will give us a glance into the customization opportunities of those elements. We will explore ways to customize those elements further in *Chapter 12, Customizing Plot Attributes – Axes, Legends, and Colors.*

In this chapter, we're going to cover the following main topics:

- The anatomy of a Plots plot
- Knowing the components of Makie's figures
- Exploring Gadfly's customizable components

Technical requirements

You will need Julia, Pluto, and a web browser with an internet connection to run the code examples in this chapter. The corresponding Pluto notebooks and their HTML versions, along with their outputs, can be found in the Chapter10 folder of this book's GitHub repository: https://github.com/PacktPublishing/Interactive-Visualization-and-Plotting-with-Julia.

The anatomy of a Plots plot

This section will describe the fundamental elements that build a figure from the Plots package. The main terminology will be helpful to us in future sections of this chapter, where we will expand on these concepts to understand Makie and Gadfly figures.

A plot from the Plots package has several components we can customize using attributes. We can think of a Plots figure as a plot that can contain subplots. Those subplots will have axes that define the plot area in which we can plot our series. We can customize each of those elements through different groups of attributes. Some attributes, especially background and foreground colors, pass their values across levels. For example, by default, the axis color, foreground_color_axis, matches the color defined for the subplot, foreground_color_subplot, which, in turn, matches the one defined for the plot through the foreground_color attribute.

Let's explore the different parts that build a Plots figure using a Pluto notebook:

1. Execute the following code in the first cell:

```
begin
    using Plots
    import Plots: mm
    default(dpi=500)
end
```

The preceding code loads the Plots package and sets the default dpi to use. It also brings the millimeters, mm, unit into scope as this will be helpful later for defining the margin sizes. Let's start exploring the axes.

2. In a new cell, execute the following code:

```
plt_axes = plot()
```

The preceding code creates an empty plot where only the axes are visible. The empty bidimensional plot has two axes, x and y, which define our coordinate system. By default, those axes go from 0.0 to 1.0, but we are free to change the axes limits using the `lims` plot attribute. The axes also have a scale. In this case, the plot has the default linear scale. The axes are essential as they give meaning to the position of our markers and lines in the plot. Therefore, we plot our data series in the plot area defined by the axes. Let's annotate these axes to visualize their parts.

3. Execute the following code in a new cell:

```
plot!(plt_axes,
    xguide = "x axis guide (label)",
    yguide = "y axis guide (label)",
    lims = (0, 1),
    ticks = 0:0.2:1,
    minorticks = 2,
    minorgrid = true)
```

The `plot!` function is modifying the plot stored in the `plt_axes` variable. It uses the `guide` attribute to assign the axis labels. Usually, we can prefix axis attributes with x, y, or z to indicate the axis we want to adjust. For example, `xguide` sets the guide or label for the *x* axis. After that, we used the `lims` attribute to specify the axis limits and ensure that the axes go from 0 to 1, independent of the data we plot.

Then, we determined the position of the axis ticks using the `ticks` attributes. The ticks are short lines perpendicular to the axis that determine the location of particular values, indicated using tick labels, in the axis. We can have smaller lines without labels showing some locations between ticks, known as minor ticks. The `Plots` package doesn't show minor ticks by default, so we used the `minorticks` attribute to divide the space between ticks by 2.

Finally, as `Plots` doesn't show the minor grid that corresponds to the minor ticks by default, we set the `minorgrid` attribute to `true` to see it. Now, let's label some of these plot elements.

4. Execute the following code in a new cell:

```
axis_annotations = [
    (0.2, 0.2, ("tick and tick label", :bottom, 10)),
    (0.5, 0.1, ("minor tick", :bottom, 10)),
    (0.75, 0.3, ("x axis border (spine)",
        :bottom, 10)),
    (0.15, 0.75, ("grid", :right, :bottom, 10)),
    (0.55, 0.65, ("minor grid", :left, :bottom, 10))]
```

The preceding code creates a vector of annotations that we will add to our plot in the next step. Each annotation tuple contains the *x* and *y* coordinates and a tuple with the annotation text and some arguments. `Plots` will pass those arguments to the `font` function to set the aspect of the annotation. In this case, we set the `valign` attribute to `:bottom` and `pointsize` to `10`. Because `Plots` uses magic arguments, we didn't need to write the names of the attributes.

5. Run the following code in a new cell:

```
plot!(plt_axes,
      [0.2, 0.2, NaN, 0.5, 0.5, NaN, 0.75, 0.75, NaN,
          0.15, 0.2, NaN, 0.55, 0.5],
      [0.2, 0.02, NaN, 0.1, 0.02, NaN, 0.3, 0.02, NaN,
          0.75, 0.6, NaN, 0.65, 0.5],
      arrow = arrow(:closed),
      color = :black,
      legend = false,
      annotations = axis_annotations)
```

The preceding code modifies our plot by adding arrows with annotations. The first two vectors we pass as the second and third positional arguments are the *x* and *y* coordinates. The first coordinates are the origin of the arrow and the second are their destination; then, a NaN value separates the arrow segments. Then, we set the `arrow` attribute to `arrow(:closed)` to add the arrowheads. Finally, we set the `annotations` keyword argument using the vector defined in the previous step to add the annotation texts. The following diagram shows the plot that was created with the axis components annotated:

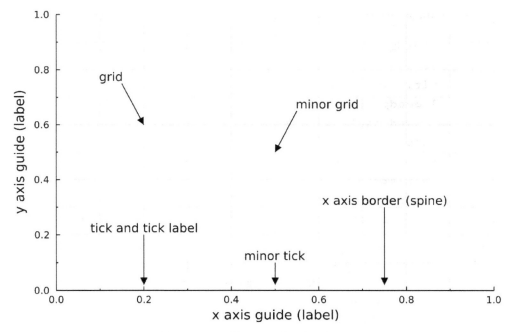

Figure 10.1 – The axes of a Plots figure

Here, we can see the different parts of the axes. The axes determine the space where we will plot our data. They use different components to help us determine positions in our coordinate system. The axis labels, or guides in Plots parlance, usually provide information about the data displayed in that dimension.

Then, the spines, or axis borders in Plots, highlight the axis directions and provide support for the ticks. The ticks and ticks labels allow us to identify particular values along the axis. The grid expands under the plot while following the ticks, enabling us to locate those positions in the plot space. Finally, we can have smaller ticks without labels, known as minor ticks, and their corresponding minor grid. Now that we have seen the parts that create Plots axes, let's explore the ones that make a subplot.

6. In a new cell, execute the following code:

    ```
    x_values = 0:0.25:1
    ```

    The preceding code creates the x coordinates for the points we will use as an example.

7. Run the following code block in a new cell:

```
plt_subplot = scatter(x_values, x_values.^2,
    zcolor = x_values, seriescolor = :grays,
    title = "Title",
    legendposition = (:topleft),
    legend_title = "Legend title",
    label = "label",
    colorbar_title = "Color bar title",
    leftmargin = 15mm,
    annotation = (0.51, 0.26,
        ("annotation text", :bottom, :left, 10)))
```

Here, we created a scatter plot for this subplot example. We can think of the subplot as a single plot containing its axes. The marker fill color indicates the content of the x_values variable, as indicated using the zcolor attribute. We used the title attribute to set the subplot title. A subplot can have a legend; we used legendposition to determine its location and legend_title to add a legend title.

We indicated the label of each series using the label attribute. In cases like this, where we have a color gradient, we can have a color bar. The color bar, like the axes, has ticks and ticks labels to indicate the scale in the color gradient. We added a color bar title using the colorbar_title attribute. In this example, we modified the left margin of the leftmargin keyword argument; you can also adjust other subplot margins.

Finally, we added an annotation text using the annotations keyword argument, as seen in *step 4 and step 5*. Let's add some annotations to highlight some of those elements.

8. In a new cell, execute the following code block:

```
subplot_annotations = [
    (0.95, 0.08, ("plot area", :right, 10)),
    (0.90, 0.71, ("color bar tick and label",
        :right, 10)),
    (0.55, 0.93, ("legend", :left, 10))]
```

As demonstrated in *Step 4*, the preceding code defines the annotations we want to add to our subplot.

9. In a new cell, execute the following code block:

```
plot!(plt_subplot,
      [0.9, 1.0, NaN, 0.55, 0.40],
       [0.71, 0.71, NaN, 0.93, 0.93],
      label = nothing,
      arrow = arrow(:closed),
      color = :black,
      background_color_inside = :gray95,
      annotations = subplot_annotations)
```

The preceding code creates arrows and annotations, as demonstrated in *Step 5*. We also set the background color of the plot area to :gray95 using the background_color_inside attribute. The following diagram shows the final plot displaying the different components of a Plots subplot:

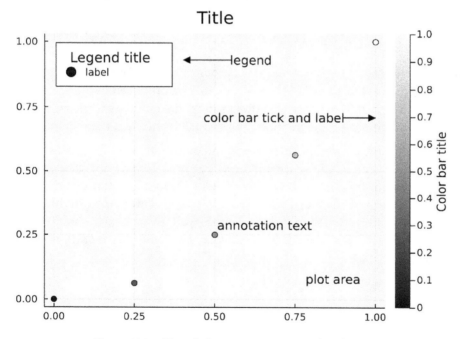

Figure 10.2 – The subplot components in a Plots figure

10. In a new cell, run the following code:

```
subplot_a = plot(title = "Subplot title")
```

Here, we created one of the two subplots that will make up our figure. We have only added a subplot title to our subplot.

11. In a new cell, run the following code:

```
subplot_b = plot(0:0.01:1, x -> x^2,
    title = "Subplot title",
    legend_position = :none)
```

Here, we created our second subplot and added a series to its axes. We have avoided showing the legend of the subplot by setting legend_position to :none.

12. Run the following code in a new cell:

```
plt = plot(subplot_a, subplot_b,
    plot_title = "Plot title",
    plot_titlevspan = 0.1,
    background_color_inside = :gray95,
    background_color = :gray85,
    layout = grid(1, 2, widths=[0.3, 0.7]))
```

Here, we used the plot function to create our plot, which contains multiple subplots. In this case, these are the two subplots we produced in the previous step. The plot_title attribute allows us to set the plot title, while plot_titlevspan indicates the fraction of the vertical space that the plot title will use.

Then, we set the background color inside and outside the plot area using the background_color_inside and background_color attributes, respectively. As a plot contains multiple subplots, it includes a layout to determine their sizes and positions. We can define it using the layout keyword argument. In this example, we used the grid function to create the grid layout to arrange the subplots. We will discuss this function in more detail in *Chapter 11, Defining Plot Layouts to Create Figure Panels*.

13. In a new cell, run the following code:

```
lens!(plt, [0.0, 0.2], [0.0, 0.1],
    background_color_inside = :gray95,
    inset_subplots = (2,
        bbox(0.25, 0.25, 0.35, 0.2)))
```

We can add inset or floating subplots inside a `Plots` plot. In this example, we used the `lens` function to add an inset plot to magnify a region of our subplot. The lens magnifies the rectangular area determined by the *x* and *y* intervals. We used two element vectors to indicate those intervals. Then, we used the `inset_subplots` attribute to determine the position and size of the floating plot. The first element of the tuple we used to set this attribute indicates the number of the parent subplot.

Here, for example, we added an inset subplot relative to the second subplot of our plot. Then, the second element of the tuple is the bounding box, `bbox`, which contains the plot area of the subplot. The two first floating points of the `bbox` function determine the location of the top-left point of our subplot. Those numbers indicate fractions in the *x* and *y* dimensions of the plot area of the parent subplot. *0.0, 0.0* is located at the top-left point in our subplot area. The last two floating-point numbers determine the inset plot's width and height as a fraction of the parent plot area. We will discuss inset plots in more detail in *Chapter 11, Defining Plot Layouts to Create Figure Panels*.

14. In a new cell, execute the following code:

```
plot_annotations = [
    (0.15, 0.9, ("bounding box", :bottom, 10)),
    (0.5, 0.85, ("inset subplot", :left, 10))]
```

The preceding code creates the annotation labels for our plot.

15. Execute the following code in a new cell:

```
plot!(plt[2],
    [0.15, 0.23, NaN, 0.5, 0.4],
    [0.90, 0.77, NaN, 0.85, 0.77],
    label = nothing,
    arrow = arrow(:closed),
    color = :black,
    annotations = plot_annotations)
```

The preceding code adds the arrows and annotations to our second subplot, as indicated by `plt[2]`. The following diagram shows the plot that was created:

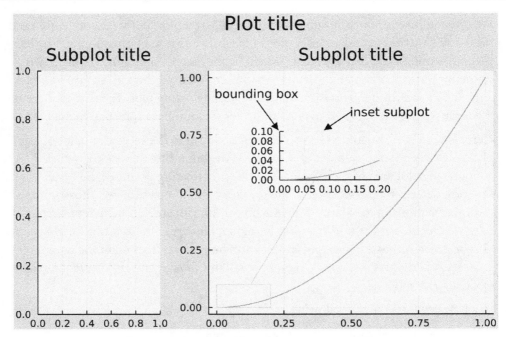

Figure 10.3 – Plots plot components

The preceding plot contains two subplots and a lens. The inset plots of the lens show a magnified region of our function near the origin. The first arrow indicates the point that determines the location of the bounding box of the inset subplot.

In this section, we learned about the main components of a Plots figure. In the next section, we will mention their Makie equivalents and discuss the differences.

Knowing the components of Makie's figures

In Makie, we can think of the Figure object as the main plot element. It contains a GridLayout that will determine the location of the different plot components in the figure. Therefore, Makie's Figure is similar to the plot object from Plots. The pieces we can arrange on GridLayout are called **layoutables**. We can find Axis, Label, Legend, and Colorbar among the many layoutables available. While Plots locates most of those elements in the subplots, Makie gives us the freedom to arrange them in Figure. We will learn more about how to place layoutables in Makie's GridLayout in *Chapter 11, Defining Plot Layouts to Create Figure Panels*.

The Axis layoutable object contains the axes for our plot, and we plot our data in it. Makie axes components are similar to those of the Plots package, which we saw in the previous section. Axis objects include the spines and the decorations: the *x*- and *y*-axis labels, the ticks and grid, and the minor ticks and the minor grid. They can also have a title similar to the subplot title of `Plots`. To have a title for the whole figure, we need to use a `Label` object. Plots' color bar title is named label in Makie parlance. The axis of the color bar is fully customizable in Makie; for example, we can modify their spines and ticks. We can also add high and low clip triangles at the extremes of the color bar.

`Figure` has its margins controlled by the amount of padding around the figure's content. The spaces between the elements in the grid layout are the gaps: `colgap` and `rowgap`. Legends also have notable margin and padding attributes. The margin is the distance between the legend border or frame and the legend bounding box. Instead, the padding is the distance between the legend content and the legend border.

Let's see an example of those Makie's plot components using Pluto:

1. Run the following code in the first cell of the notebook:

    ```
    begin
        import Pkg
        Pkg.activate(temp=true)
        Pkg.add(name="CairoMakie", version="0.8.8")
        using CairoMakie
    end
    ```

    The preceding code installs and loads `CairoMakie` 0.8.8 in a temporal environment. Note that this cell can take some time to run and that the output that's captured in Pluto's log can be large.

2. Execute the following code in a new cell:

    ```
    begin
        fig = Figure(fontsize=24)
        axis = Axis(fig[1, 1])
        plt = scatter!(axis, 1:10,
            color = 1:10,
            colorrange = (0.0, 10.0),
            label = "Legend's label")
        fig
    end
    ```

    Here, we used the `Figure` constructor to create an empty figure containing `GridLayout`. We populated the layout by adding an `Axis` object to the first grid cell. Then, we used the `scatter!` function to draw 10 points in our axes.

This function returned the plot object; we will need it in future steps. We used the `color` and `colorrange` attributes to color the points according to a color gradient. We also labeled our series using the `label` attribute – the `Legend` constructor will need it.

3. In a new cell, execute the following code:

```
begin
    legend = Legend(fig[2, 1], axis,
        "Legend's title",
        tellheight=true, tellwidth=false)
    cbar = Colorbar(fig[1, 2], plt,
        label = "Colorbar's label (title)",
        ticks = 0:1:10)
    fig
end
```

Before we ran this cell, our `Figure` constructor only carried the `Axis` object. The preceding code adds a `Legend` object for the series in our axes. We placed the legend in `GridLayout`, just under our axis. The string in the third positional argument determines the legend's title. We set the `tellheight` attribute to `true` so that the cell grid knows the height of the legend. We also set `tellwidth` to `false` to avoid the `GridLayout` column's width from matching the legend's width. Then, we added a `Colorbar` object to the right-hand side of our `Figure` constructor. We used the `label` attribute of the `Colorbar` constructor to set its title, and the `ticks` attribute to increase the number of color bar ticks.

4. In a new cell, execute the following code:

```
begin
    hidespines!(axis, :t, :r)
    axis.xminorgridvisible = true
    axis.yminorgridvisible = true
    axis.xminorticksvisible = true
    axis.yminorticksvisible = true
    axis.title = "Axis' title"
    axis.xlabel = "Axis' xlabel"
    axis.ylabel = "Axis' ylabel"
    fig
end
```

Here, we set some attributes for our Axis object. First, we hid the top and right spines that Makie shows by default using the hidespines! function. In particular, we made the minor grid and ticks for the *x* and *y axes* visible. Then, we set a title for the axes and a label for each axis. The following diagram shows the resulting plot:

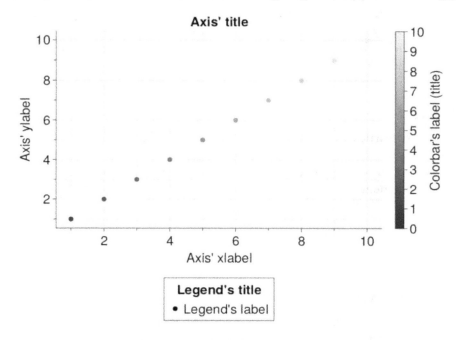

Figure 10.4 – Makie's nomenclature

Now that we have seen the main similarities and differences between Plots and Makie figures, let's compare the former and Gadfly.

Exploring Gadfly's customizable components

This section will briefly mention the most notorious differences between Plots and Gadfly plot component nomenclature. In *Chapter 5*, *Introducing the Grammar of Graphics*, we described some components of a Gadfly figure. Here, we will highlight Guides, Coordinates, and Scales. Scales determines the scales used by the axes, color, sizes, shapes, and line styles. Coordinates chooses the coordinate system for our axes; however, in Gadfly 1.3, only cartesian coordinates are available.

Finally, Guides offers support for adding and customizing axes, annotations, titles, and keys. The keys generate the legends of the plot; we can have color, shape, and size keys. If the color scale is continuous, the color key will create a color bar rather than something similar to a `Plots` legend. Gadfly's `Guides` will allow you to modify the axes ticks and labels and add rugs. These rugs are short lines that indicate the position in the axes where there is a data point.

`Gadfly` generally allows for fewer customization opportunities compared to `Plots` and Makie. Therefore, `Gadfly` needs fewer definitions for its plot components. However, as we can see, `Gadfly` names its elements similar to `Plots`, but it organizes them differently. We quickly discussed this organization here to allow you to find the attributes for customizing the components.

Summary

In this chapter, we learned about the main components of a figure from the `Plots` package. Then, we briefly extended that description to Makie and `Gadfly` figures. The knowledge we've acquired in this chapter will help us when we look for attributes to customize those elements. We also gained an idea about the place that the layout takes in `Plots` and Makie figures.

We will describe the layout capabilities of those packages in more detail in the next chapter.

Further reading

The knowledge you've acquired in this chapter will help you start exploring how to customize the plot components while following the documentation for `Plots`, Makie, and `Gadfly`. In the case of `Plots` and Makie, you can explore some of those attributes in *Chapter 12, Customizing Plot Attributes – Axes, Legends, and Colors*. For more information about the layout capabilities, please refer to *Chapter 11, Defining Plot Layouts to Create Figure Panels*.

11
Defining Plot Layouts to Create Figure Panels

Sometimes, we need to create figures or plot panels that contain a set of related plots. For example, when making figures for publication in scientific journals, there is usually a constraint on the size and number of figures. In those cases, we need to decide the figure's layout, determining the placement and relative sizes of the contained plots. A common practice is to export the plots in *SVG* format and then use a vector graphics editor, such as *Inkscape*, to define the figure's layout. However, that practice hampers reproducibility, as changes to the included plots can require us to manually redo the figure. Thankfully, `Plots` and *Makie* allow us to layout our figures using Julia code rather than external programs, ensuring reproducibility and easing the figure creation process.

In this chapter, we will learn to compose multiple plots to create complex figures using `Plots` and *Makie*. These packages provide excellent utilities to place our plots into a grid with custom column and row sizes effortlessly. However, their layout system is a lot more flexible and powerful; you can, for example, nest layouts or place an object at any desired place in the figure. Besides creating static figures, *Makie's* layout system also allows us to control the placement of interactive widgets, such as sliders and buttons. At the end of the chapter, we will also briefly discuss the functions offered by `Gadfly` for composing plot panels.

In this chapter, we're going to cover the following main topics:

- Creating layouts with Plots
- Understanding Makie's layout system
- Composing Gadfly plots

Technical requirements

You will need Julia, Pluto, and a web browser with an internet connection to run the code examples. The corresponding Pluto notebooks and their HTML versions with outputs are in the Chapter11 folder of the book's GitHub repository: https://github.com/PacktPublishing/Interactive-Visualization-and-Plotting-with-Julia.

Creating layouts with Plots

We already saw a little introduction to layouts using Plots in the *Simple layouts* section of *Chapter 1, An Introduction to Julia for Data Visualization and Analysis*. In particular, we have seen that Plots automatically composes a figure when we pass multiple plots into the plot function. We have learned how to gain more control over the final subplot placement using the layout attribute of the plot function. To help us arrange the subplots, we saw how to create a grid layout using the grid function. We also learned to use the grid function's widths and heights keyword arguments. These arguments define the relative proportion of the final figure assigned to each cell of the grid layout. Finally, in that section, we saw that we can use the link keyword argument of plot to match the axis across subplots.

Then, we introduced more complex layout topics in *The anatomy of a Plots plot* section of *Chapter 10, The Anatomy of a Plot*. Specifically, we introduced inset plots, specified using the inset_subplots attribute of the plot function, and bounding boxes, created using the bbox function.

In this section, we will further explore those functions and present a different way to create plot layouts with Plots. In particular, we will introduce the @layout macro, which allows us to express plot layouts using the Julia syntax for matrix creation. Let's create a new Pluto notebook to explore how to generate complex plot arrangements:

1. Execute the following code in the first cell of the notebook:

```
using Plots, PlutoUI
```

This loads the `Plots` package for plotting and `PlutoUI` for creating the interactive sliders that we will use to explore how changes in some arguments affect the final layouts.

2. Create a new cell and run the following code:

```
@bind fraction Slider(0.1:0.1:0.9,
    default=0.4, show_value=true)
```

This creates a slider to set and modify the fraction of the width and height of the final figure that the first subplot will occupy. The slider can set values between 0.1 and 0.9 to avoid creating a subplot with zero dimensions or covering the whole figure.

3. Execute the following code in a new cell:

```
layout_grid = grid(2, 2,
    heights = [fraction, 1.0 - fraction],
    widths = [fraction, 1.0 - fraction])
```

We use the `grid` function to create a grid layout with two rows and two columns. We set the first row's height and the first column's width using the `fraction` value defined in the previous step. The sum of the widths and heights should not be greater than 1.0. Therefore, we determine the second row and second column sizes as the remaining fraction.

4. In a new cell, run the following code:

```
plot(plot(log), plot(exp),
    plot(sqrt), plot(x -> x^2),
    legend = false,
    layout = layout_grid)
```

We create a figure with four plots, and we set the `layout` attribute using the grid layout defined in the previous step. The first subplot uses 40% of the figure height and width, as indicated by the default 0.4 value that we set in *step 2*. You can modify the proportions by moving the slider defined in that step. The following figure shows the created plot:

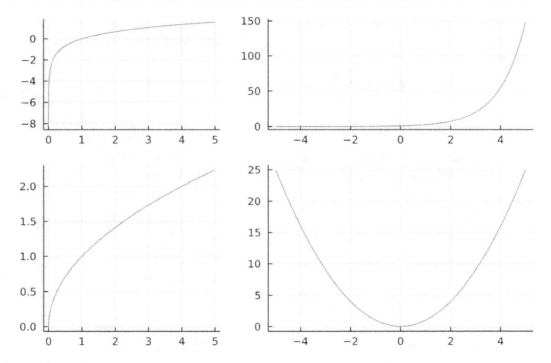

Figure 11.1 – The Plots figure containing four subplots

5. Run using Plots.PlotMeasures in a new cell to use the w, figure width, and h, figure height, plot measures; we will need it later to set the subplot sizes in the @ layout macro.

6. In a new cell, execute the following code:

```
layout_macro = @layout [ a{0.4h, 0.4w} b
                         c                 d ]
```

We create a figure layout using the @layout macro and the Julia matrix syntax, composed of squared brackets, spaces, and newlines or semicolons. In this example, we create a matrix with two rows and two columns. Each matrix value indicates the label that Plots will assign to each subplot in the layout; you don't need to do anything with that value. We can use any valid identifier as a label. However, there is a label with a special meaning – the underscore (_). As we will see later in this section, it allows us to leave empty spaces in our figure. In this case, we have used a, b, c, and d.

Finally, we can use braces after the subplot label to indicate the desired size. If we do not use braces, @layout divides the area of the figure into equal parts. In this example, we set the first subplot to have 40% of the figure height, as indicated by 0.4h, and 40% of the width, indicated by 0.4w. We have set both dimensions, but we can select only one. In that last case, Plots divides the remaining dimension into equal parts. Let's now create a figure using this layout.

7.   In a new cell, execute the following code:

```
plot(plot(log), plot(exp),
    plot(sqrt), plot(x -> x^2),
    legend = false,
    layout = layout_macro)
```

This creates a figure using the layout we created with the @layout macro. The constructed plot is identical to the one displayed in *Figure 11.1*, where the first row and column take 40% (0.4) of the figure's total height and width respectively.

8.   In a new cell, execute the following code:

```
plot(plot(log), plot(exp),
    plot(sqrt), plot(x -> x^2),
    legend = false,
    layout = @layout [ a grid(3, 1) ])
```

We create a *nested layout* using the @layout macro in this example. The @layout macro defines an arrangement with one row and two columns using the Julia matrix syntax. But, in the second cell, we include a grid layout with three rows and one column, created using the grid function. We use grid, but we can also nest calls to the @layout macro. The following figure shows the plot created by this command:

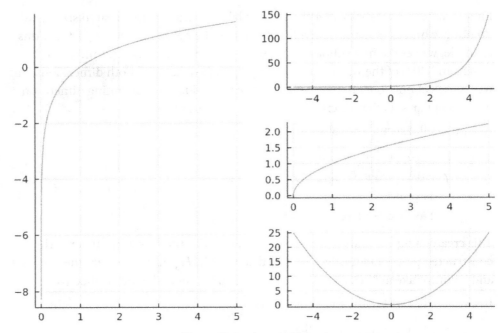

Figure 11.2 – A nested layout

9. In a new cell, execute the following code:

```
layout_macro_blank = @layout [ a b _
                               _ c d ]
```

Here, we create a layout using the @layout macro and the particular behavior of _ as a subplot label. The designed arrangement has two rows and three columns but only four subplots instead of six. The cell named _ will be empty in the final figure. Let's create a plot using this layout to see it in action.

10. In a new cell, execute the following code:

```
plot(plot(log), plot(exp),
     plot(sqrt), plot(x -> x^2),
     legend = false,
     layout = layout_macro_blank)
```

This code creates a figure containing the four subplots but six cells. We indicate the layout places that remained blank using _ in the @layout macro of the previous step. The following figure shows the created plot:

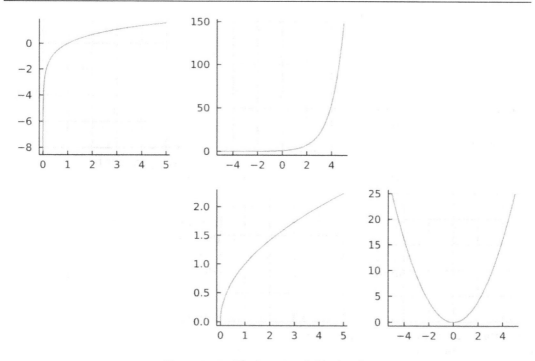

Figure 11.3 – The layout with blank cells

11. In a new cell, execute the following code:

```
@bind pos Slider(0.0:0.01:0.6,
        default=0.13, show_value=true)
```

This code defines a slider to set the pos variable that we will use to set the position in the *x* and *y* dimensions of our inset plots. We stopped at 0.6 to avoid the inset plots from going outside our canvas.

12. Run the following code block in a new cell:

```
begin
    plt = plot(
        plot(x -> -x), plot(x -> -x),
        legend = false)
    plot!(plt, exp,
        inset = bbox(pos, pos, 0.2, 0.2),
        subplot = 3, legend = false)
    plot!(plt, log,
        inset = (2, bbox(pos, pos, 2.5cm, 2.5cm)),
```

```
        subplot = 4, legend = false)
end
```

The first call to `plot` creates a figure containing two subplots. By default, `Plots` created a layout with one row and two columns, each with the same width. We assign the resulting plot into the `plt` variable to later modify it using the `plot!` function. Each call to `plot!` will create an *inset plot*. The first will be bound to the whole figure, while the second will be inside the plot area of the second subplot. For that, we will use the `inset` and `subplot` attributes. The `inset` attribute will define the bounding box for the inset plot and its parent. The `subplot` attribute will indicate the subplot where the `plot!` function will plot the series.

In the first call to `plot!`, we create an inset plot with the whole figure as its parent. We achieve this by assigning only the bounding box to the `inset` attribute. In the second call, we pass a two-element tuple, where the first element indicates the subplot index of the parent and the second the bounding box. We create the bounding boxes using the `bbox` function.

The first positional argument of the `bbox` function determines the position in the *x* dimension as a fraction of the parent element. The second positional argument sets the *y* dimension. In this case, we have specified both using the value of the `pos` variable defined in the previous step with the slider. You can move the slider to get a better sense of the meaning of those positions. As you will see, the origin is located in the top-left corner of the parent area by default. We can change that by setting the last position arguments or the `h_anchor` and `v_anchor` keyword arguments of the `bbox` function. We can set `h_anchor` to `:left`, `:right`, and `:hcenter` and `v_anchor` to `:top`, `:bottom`, and `:vcenter`. If we pass any of those values as the fifth and sixth positional arguments, `bbox` will assign them to the corresponding dimension. We can also pass `:center` as the fifth positional argument to center in both dimensions.

The third and fourth positional arguments of the `bbox` function determine the bounding box's size. In particular, the third one determines the width and the fourth the height. If we set those positional arguments using floating-point values, `Plots` will interpret them as a fraction of the parent. For example, the first call to `bbox` creates a bounding box with 20% (`0.2`) of the width and height of the whole figure. The second, instead, uses the `cm` unit exported by the `PlotMeasures` module of `Plots` that we loaded in *step 5*. So, in this case, we created a squared inset plot with `2.5` centimeters on each side. We have used `cm`, but we can use any of the module's other units, such as `inch`, `mm`, `pt`, or `px`.

The following figure shows the created plot before interacting with the slider:

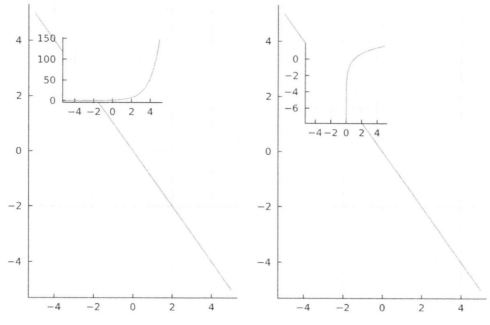

Figure 11.4 – Inset plots

These examples show you how to create a layout using the `grid` function and the `@layout` macro. We can use these to create complex layouts. For example, we can use the `@layout` macro to create a nested layout by including a layout created with `grid` or `@layout` inside the matrix. In the last example, we also saw how we can use inset plots to go beyond grid layouts.

In the following section, we will discuss the creation of figure layouts using Makie.

Understanding Makie's layout system

We have already used some of Makie's layout capabilities in the *Interactive and reactive plots with Makie* section of *Chapter 3, Getting Interactive Plots with Julia*. In this section, we will further discuss them. As we mentioned in the *Knowing the components of Makie's figures* section of *Chapter 10, The Anatomy of a Plot*, a `Figure` object contains a `GridLayout` to indicate how to place the different layoutables in the final figure. Makie layoutables can be subplots, as in the case of `Plots`, or even other objects such as sliders and text, as shown in the example from *Chapter 3, Getting Interactive Plots with Julia*. Makie's design gives us a lot of freedom when conceiving the arrangement of the different parts of the figure.

We can assign elements to one or multiple cells of a grid layout using the indexing syntax. For example, indexing a `Figure` object will select the indicated positions for its inner grid layout. We can explicitly access the grid layout of a figure by accessing its `layout` field. If the cell we accessed using the indexing syntax doesn't exist, Makie will update the dimensions of the grid to create it. For example, plotting into `figure[2, 2]` or a newly created figure will generate a two-by-two grid layout. Then, if we plot into `figure[1:2, 1]`, the subplot will expand through the first two rows of the first column.

Makie's indexing syntax for working with the underlying layout is convenient. However, as with `Plots`, we can make more complex arrangements by nesting grid layout objects. We can explicitly create an empty layout calling `GridLayout` with the desired number of rows and columns as positional arguments. If we do not give those values, Makie will create a one-by-one grid. We can assign the designed grid layout to a variable to allow for further customization later.

When customizing a Makie layout, it is helpful to know that Makie aligns the elements through the inner plotting area by default. All the parts outside the plotting area, such as axis labels, ticks, and titles, are **protrusions**. Therefore, Makie locates those protrusions between the aligned plotting areas in the grid. Layoutable objects, such as `Axis` and `Legend`, have an `alignmode` attribute to determine how they align in the grid. The default align mode is `Inside`, causing the described behavior. We can set `alignmode` to `Outside`, aligning the protrusions, or `Mixed`. This can take the `left`, `right`, `bottom`, and `top` keyword arguments. If we do not set a side, its default mode will be `Inside`. However, assigning a `Real` number – for example, `0.0` – will make that side have an `Outside` align mode with the given number as padding.

Makie also offers functions to control the size of rows, columns, and the space between those: `rowsize!`, `colsize!`, `rowgap!`, and `colgap!`. These functions take a `GridLayout` object in their first positional argument. Then, the following positional argument is the row or column number we want to modify. After that, we can use an `Auto`, `Relative`, or `Fixed` object to set the size. The last two can take floating-point numbers indicating a specific size or a fraction relative to the parent container.

Let's create a new Pluto notebook to see in action some of the previously described behaviors for Makie's layout system:

1. Run the following code in the first cell:

    ```
    using CairoMakie
    ```

    It loads *Makie* through the `CairoMakie` backend — the best backend to produce publication-quality plots.

2. Create a new cell and execute the following code:

```
begin
    xs = 0:0.1:2pi
    fig = Figure()
    axis_a = Axis(fig[1,1])
    lines!(axis_a, xs, sin)
    fig
end
```

Here we created a new Figure and added an Axis block into the first cell of the GridLayout of the Figure using the indexing syntax. Then, we used the lines! function to plot the sin function into this Axis object.

3. Execute the following code in a new cell:

```
begin
    layout_b = GridLayout()
    fig[1,2] = layout_b
    axis_b_1 = Axis(layout_b[1,1])
    axis_b_2 = Axis(layout_b[2,1])
    scatter!(axis_b_1, xs, sin, color=sin.(xs))
    scatter!(axis_b_2, xs, cos, color=cos.(xs))
    fig
end
```

We have created a new GridLayout and stored it in the layout_b variable. Then, we assigned this GridLayout to the second cell of the figure's GridLayout to create a nested layout. In particular, we located the nested layout in the first row and the second column of the figure's layout: fig[1,2]. After that, we added two Axis blocks in the new layout; one in the first row and the other in the second one. Finally, we plot a scatter plot on each Axis using the scatter! function. As we can see, we plotted the sin function in the Axis located in the first row and the cos function in the other. Also, we used the color attribute to color each point accordingly to its *y* coordinate. Let's add a color bar for these plots in the following step.

4. In a new cell, execute the following code:

```
begin
    Colorbar(layout_b[1:2, 2], colorrange=(-1,1))
```

```
        fig
    end
```

We added the `Colorbar` in the second column of the nested grid layout across the two rows of the layout, as indicated by the `1:2` range. Since we have not linked the color bar to a plot, we manually set its `colorrange` to match the limits on the *y* axis.

5. Execute the following code in a new cell:

```
begin
    colgap!(fig.layout, 40)
    fig
end
```

This code block uses the `colgap!` function to increase the space between the two panels of the figure. Note that this function takes a layout in the first positional argument, so we passed the grid layout contained in the `layout` field of the `Figure` object.

6. In a new cell, execute the following code:

```
begin
    Label(fig[1, 1, TopLeft()], "A",
    textsize = 24, padding = (0, 35, 10, 0))
    Label(fig[1, 2, TopLeft()], "B",
    textsize = 24, padding = (0, 35, 10, 0))
    fig
end
```

Here we have added a label to each of the cells of the figure's `GridLayout` using the `Label` constructor. In particular, we added the `A` label to the first cell of the grid layout — note that we are adding two layoutables in the same cell; the `Axis` and the `Label` objects. We used `TopLeft()` as the index for the third dimension of the `GridLayout` to indicate the layoutable position. Finally, we used the `textsize` attribute to increase the default size and padding to customize the location of the labels. The `padding` attribute of `Label` takes a tuple indicating the padding on the left, right, bottom, and top sides of the bounding box containing the label, in that order. The following figure shows the final plot panel we have created:

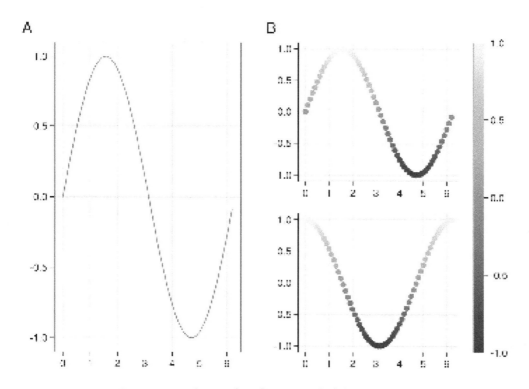

Figure 11.5 – Plot panel made using Makie's layout system

Finally, Makie can also create inset plots by plotting different subplots into the same grid location or creating a bounding box with BBox. However, the Makie syntax for creating inset plots is less convenient than the Plots syntax.

In this section, we have learned about the main functions and objects that allow us to create complex layouts with Makie. We will look at the principal functions to arrange plots with Gadfly in the following section.

Composing Gadfly plots

In the *Exploring data with Gadfly* section of *Chapter 5, Introducing the Grammar of Graphics*, we saw how to create small multiples and horizontally stack plots. For the latter, we used the hstack function, which takes a series of Gadfly plots, and stacks them horizontally, creating one row and multiple columns. We can use the vstack function to arrange the subplot vertically and create multiple rows. We can combine the result of vstack using hstack, or vice versa, to create a grid. However, it is also possible to use the gridstack function for that. The input of gridstack is a matrix of plots that indicates the position of each subplot in the grid.

In this section, we have noted the three main functions offered by `Gadfly` to arrange multiple plots into a single figure. Let's summarize what we've learned so far.

Summary

In this chapter, we have learned about the `Plots` and *Makie* mechanisms for arranging plots into complex figures. Having briefly touched on the topic in previous chapters, this one has gone deeper into the description of their layout system. In particular, we have seen more detail on how we can organize subplots using the `Plots` package. We have also expanded the concepts explored previously in creating grid layouts using *Makie* and `Gadfly`. We will find this knowledge valuable when creating plot panels and complex figures for publication in different media. In the case of *Makie*, this will also help us create interactive applications.

In the following chapter, we will learn to use the different plot attributes to customize the aspects of axes, legends, and colors.

Further reading

To better understand the layout system of Makie, we recommend exploring the documentation of the package and its layout examples and tutorial at `https://makie.juliaplots.org/stable/`.

12

Customizing Plot Attributes – Axes, Legends, and Colors

Depending on your work, you might find it helpful to customize different visual aspects of your plots and figures. For example, you might need to increase the number of axis ticks to make a plot more readable, to inform units in the axis labels, or choose color-blind friendly colors. `Plots` and *Makie* are very flexible plotting libraries, allowing you to control that and many more aspects of your figures. These libraries achieve that thanks to a large set of plotting attributes, some of which we will explore in this chapter. Moreover, we will learn to explore the available plotting attributes from our Julia session.

In this chapter, we will learn how to customize text elements, axes, legends, and colors using `Plots` and *Makie*, focusing on the former. What's more, we will explore other packages to help us add LaTeX equations to our figures and change color palettes.

In this chapter, we're going to cover the following main topics:

- Exploring plot attributes
- Using LaTeX equations
- Formatting the fonts

- Customizing the axes
- Tailoring legends
- Coloring our figures

Technical requirements

For this chapter, you will need the following:

- Julia 1.6 or higher
- `Plots` – the code for this chapter uses version 1.29.1
- Pluto version 0.18.3 or higher
- A web browser
- An internet connection

The code examples and the Pluto notebooks, including their HTML versions with embedded outputs, are in the `Chapter12` folder of the book's GitHub repository: `https://github.com/PacktPublishing/Interactive-Visualization-and-Plotting-with-Julia`.

Exploring plot attributes

`Plots` and *Makie* offer numerous attributes – keyword arguments – that we can use to determine the aspect of our figures. Describing all of them exceeds the objective of this chapter. Thankfully, both packages provide excellent ways to explore them by ourselves without leaving the *Julia REPL* or *notebook*. In this section, we will learn to use those tools using Pluto — we need Pluto version 0.18.3 or higher to see the standard output under the executed cell. Let's start examining the functions offered by the `Plots` package to explore its plotting attributes:

1. Create a new Pluto notebook and execute the following code in the first cell:

   ```
   using Plots
   ```

   Here, we load `Plots` to access its tools to inspect the different attributes. This cell can take some time to execute.

2. Create a new cell and execute the following code:

   ```
   plotattr(:Axis)
   ```

The `plotattr` function takes `Symbol`, naming a plot element, and prints the list of the attributes we can use to adjust it. In particular, we can use one of the following symbols: `:Axis`, `:Subplot`, `:Plot`, and `:Series`. We discussed those elements in the *The anatomy of a Plots plot* section of *Chapter 10, The Anatomy of a Plot*. In this case, we use the `plotattr` function to show all the available attributes to customize plot axes.

3. Execute the following code in a new cell:

    ```
    plotattr("gridlinewidth")
    ```

    When we call the `plotattr` function using an attribute's name or alias as a `String` object, it displays the attribute's documentation. That's because the `plotattr` function has two methods. One of the methods takes `Symbol`, representing a plot element, and the other takes `String`, representing an attribute. Here, we request the documentation of the `gridlinewidth` attribute, previously listed in the attributes for `Axis`. We can see at the end of the following figure the documentation printed by the `plotattr` function to the standard output:

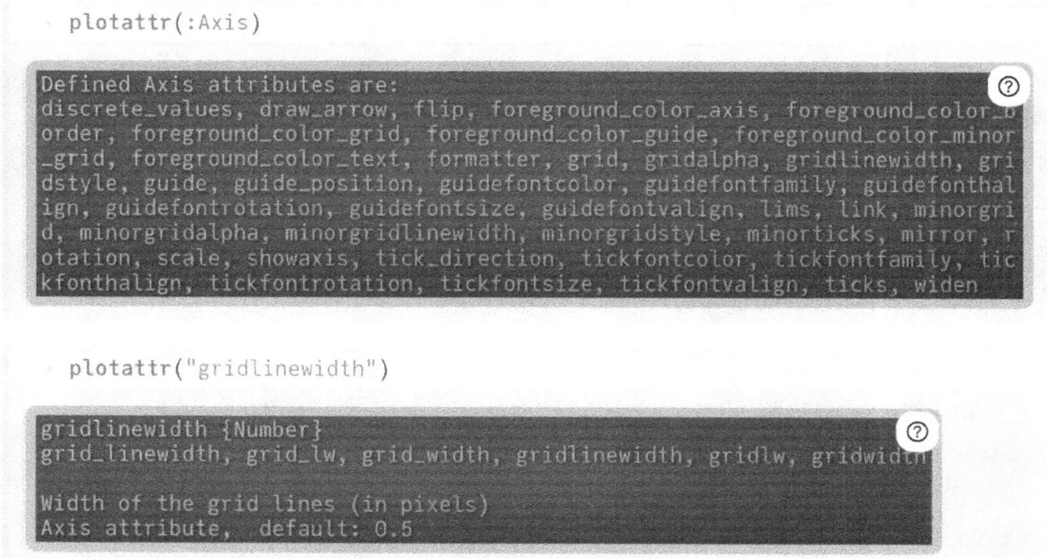

    ```
    plotattr(:Axis)
    ```

    ```
    plotattr("gridlinewidth")
    ```

Figure 12.1 – The attribute documentation in Plots

Let's analyze the printed documentation for the `gridlinewidth` attribute. It starts by naming the requested attribute and then shows the data types accepted by this attribute between curly brackets. For example, we can only set the `gridlinewidth` attribute using `Number`.

On the second line, we have the list of aliases available for this attribute. In this case, we can set this attribute using any of the following keyword arguments: `gridlinewidth`, `grid_linewidth`, `grid_lw`, `grid_width`, `gridlw`, and `gridwidth`. Using aliases is helpful when quickly testing ideas. Still, we recommend using the real attribute names to avoid confusion and make it easier to understand the code. The paragraph after the attribute name and its synonyms shows the attribute's description.

Finally, we have text indicating the element to which this attribute belongs and its default value. In this case, it is an `Axis` attribute that `Plots` sets to `0.5` by default.

4. In a new cell, execute the following code:

```
default(:gridlinewidth)
```

The `default` function takes a `Symbol` instance containing the attribute name or its alias and returns the default for it. For example, the executed code returns `0.5`. You can also use the `default` function to change the default value of an attribute by passing an extra position argument or using keyword arguments.

So, we saw that `plotattr` is our primary function to explore attributes in `Plots`. Let's now create a new Pluto notebook to explore attributes in *Makie*:

5. Create a new Pluto notebook and execute the following code in the first cell:

```
using CairoMakie
```

Here, we load *Makie* using the `CairoMakie` backend, but you are free to select another if you prefer. This cell will run for some time the first time you run it, as it is installing and precompiling many Julia packages.

6. Click on the **Live docs** button – the one with the *books* emoji – at the bottom right of the Pluto notebook, just above the footer. This will open the *documentation panel* shown in the following screenshot:

Figure 12.2 – Pluto's Live docs documentation panel

7. Type `Axis` in the **Search docs...** input located next to the title of the documentation panel, just after the *open book* emoji shown in *Figure 12.2*. After some seconds – it can take some time the first time you use it – you will see the documentation for the `Axis` object. *Makie* locates the attribute list inside the object's documentation.

8. Press *Ctrl + F* to open the search box of your browser.

9. Type `grid` in the search box; you will see all the words that contain *grid* highlighted in the documentation. That way, you can easily navigate all the attributes related to the customization of `Axis` grids. In the following figure, we can see an example of this:

Figure 12.3 – An attribute description in Makie's documentation

We saw the attribute description in the previous steps using the Pluto documentation panel. Suppose you are at the *Julia REPL* instead. In that case, you can access the documentation of the desired *Makie* object using the *help mode* of the REPL. You can see an example of how to use it in the *REPL modes* section of *Chapter 1, An Introduction to Julia for Data Visualization and Analysis*. As this chapter focuses mainly on Plots, we will hint at the related Makie objects in each section. Then, you can easily explore the attributes of those objects from Julia.

Now that we know how to explore Plots and Makie attributes to customize our figures without going out from Julia, let's discuss some exciting customization for these packages.

Using LaTeX equations

When plotting with `Plots` and *Makie*, we can use LaTeX equations created with the `LaTeXStrings` package. In particular, we can use the *L string macro*, `@L_str`, to create such equations. When using `Plots`, we will need to load the `LaTeXStrings` package to use the `L` string macro. Loading that package is unnecessary when using Makie, as Makie automatically exports that macro. The syntax is simple; you only need to write your LaTeX equation inside a string and add an `L` prefix before the *quotation mark*, as shown in the following line of code:

```
L"x^2 + y^2 = 1"
```

When we create a `LaTeXString` object using the L string macro, as in the previous code, we do not need to enclose the equation within the dollar symbols. However, you can use them to add an equation inside a longer string, as shown in the following example:

```
L"Unit circle: $x^2 + y^2 = 1$"
```

As the dollar symbol, $, has a special meaning inside LaTeX strings, we will need to use `%$` to interpolate values inside them. For example, this can be useful when creating LaTeX labels for the x-axis ticks:

```
[ L"%$i" for i in -1:0.5:1 ]
```

The previous code used array comprehension to create a vector of LaTeX strings, each containing a number between -1 and 1.

We can use LaTeX strings for any text in a plot, such as axis labels and annotations. For example, the following code will create a figure in which all the text elements are LaTeX strings:

```
using Plots, LaTeXStrings

plot(cos.(0:0.01:2pi), sin.(0:0.01:2pi), ratio = :equal,
    label = L"Unit circle: $x^2 + y^2 = 1$",
    annotation = (0, 0, L"(0,0)", 10),
    xlabel = L"x",
    ylabel = L"y",
    xticks = (-1:0.5:1, [ L"%$i" for i in -1:0.5:1 ]),
    yticks = (-1:0.5:1, [ L"%$i" for i in -1:0.5:1 ]))
```

In this example, we have used `LaTeXString` equations for the x and y labels and the annotation, an equation inside a string for the label, and we interpolated a numerical value for the x and y ticks. The following figure shows the created plot:

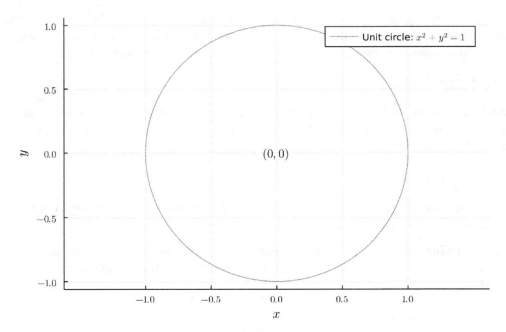

Figure 12.4 – Text elements using LaTeXStrings

To use LaTeX strings with the `Plotly` and `PlotlyJS` backends, you need to set the `extra_plot_kwargs` attribute of `plot` to specify the `include_mathjax` parameter. We can use the `MathJax` library from a *CDN*, through the `"cdn"` option, or a user-defined file. For example, we can set the attribute in the following way to make the previous example work with Plotly-based backends:

```
extra_plot_kwargs = KW(:include_mathjax => "cdn")
```

As with `Plots`, we can set text elements in Makie using LaTeX strings created with the L string macro. `Gadfly`, instead, doesn't support `LaTeXString` objects. To use some LaTeX equations with `Gadfly`, we can use the `to_latex` function from the `UnicodeFun` package to translate a string containing a LaTeX equation to its Unicode equivalent, as shown in the following example:

```
using UnicodeFun
to_latex("x^2 + y^2 = 1")
```

In this section, we have learned to use LaTeX strings to add equations to our figures. We have used it, for example, to customize our axis ticks and labels. In the following section, we will learn to customize other text-related aspects further.

Formatting the fonts

In the previous section, we saw that we can use `LaTeXString` objects from the `LaTeXStrings` package to include LaTeX equations in our plots. In this section, we will go deeper into the formatting of the different text elements of our figures. Therefore, we will learn how to customize the `Font` objects using the `Plots` package. We can create these objects using the `font` function. The `font` function takes the following keyword arguments to define the different aspects of the font:

- `family`: This takes a string defining the font family. The possible values are `"serif"`, `"sans-serif"`, and `"monospace"`. By default, Plots uses a sans serif font.

- `pointsize`: This takes an integer defining the font size in points.

- `color`: This takes a `Colorant` object from the `Colors` package or a `Symbol`, indicating the color name to define the color of the text.

- `halign`: This defines the horizontal alignment of the text using one of the following symbols: `:hcenter`, `:left`, and `:right`. By default, Plots centers the text on the horizontal axis.

- `valign`: This takes a symbol between `:vcenter`, `:top`, and `:bottom` to determine the vertical alignment of the text. Plots centers the text in the vertical dimension by default.

- `rotation`: This takes a floating-point number, indicating the rotation angle for text in degrees.

Then, `Plots` offers attributes to set those properties for the different text elements. Those attributes have a common *suffix*, starting with `font` and ending with the name of the keyword argument in the `font` function – for example, `fontcolor`. There is an exception for `pointsize`, as the corresponding suffix is `fontsize`. The *prefix* of the attribute indicates the element we want to modify. For example, we can set the color of the axis guide text using the `guidefontcolor` attribute. The available prefixes are as follows:

- `guide`: To determine the font aspect of the axis labels

- `tick`: To modify the appearance of the axis tick labels

- `colorbar_tick`: To set the text aspect of the ticks labels in the color bar

- `colorbar_title`: To set the esthetic characteristics of the font used for the title of the color bar

- `title`: To determine the aspect of the subplot title

- `plot_title`: To determine the visual aspect of the plot title
- `legend`: To set the font attributes of the item labels inside the legend
- `legendtitle`: To set the font parameter for the legend title

For example, changing the font family for the legend title can be achieved by setting the `legendtitlefontfamily` attribute to `"serif"`. We can also set the same attributes for annotations using the magic arguments during their definition, as we did in *Chapter 10, The Anatomy of a Plot*. However, Plots also offers the `annotationfontfamily` and `annotationfontsize` attributes for determining the aspect of the annotations.

Now, we know how we can change the aspect of the different text elements in our figures. That can be helpful when creating figures for publication, needing a particular font family or size. In the following section, we will discuss some Plots attributes to customize the axis, besides the attributes of their text elements.

Customizing the axes

In the previous section, we saw how to customize the text elements of the axes. In this section, we will learn about the `Plots` attributes to customize other aspects of the axes, except those determining the colors, as we will discuss them later in this chapter. Look at the `Axis` type documentation to gain insight into the *Makie* attributes to customize 2D axes.

All the `Plots` attributes we mention in this section modify all axes simultaneously. You need to use the x, y, or z prefix to alter the attribute of a particular axis. For example, `scale=:log10` will make all axes have a logarithmic scale, while `xscale=:log10` will set such a scale only for the *x* axis. Among the attributes specifying the axis aspects, we can find the following:

- `scale`: By default, Plots sets this attribute to `:identity`, to use the standard linear scale. Then, depending on the backend, other scales are supported. To use a logarithmic scale, you can choose from `:ln`, `:log2`, and `:log10`.

- `flip`: This takes a Boolean value, `false` by default, indicating whether Plots should reverse the axes. You can use the `xflip` and `yflip` attributes if you only want to flip the *x* or *y* axis respectively. We showcased the `yflip` attribute in the *Heatmap* section of *Chapter 1, An Introduction to Julia for Data Visualization and Analysis*.

- `formatter`: We can set this to `:scientific`, `:plain`, or `:auto`, the default. It determines the method used to convert the tick positions into the strings used to label them. Moreover, we can also set this attribute using a *custom function* to convert numbers into strings.

- `draw_arrow`: This takes a Boolean value, indicating whether we want to draw arrows at the axis extremes. By default, Plots set this attribute to `false`, avoiding the arrowheads at the axis ends. As of Plots version 1.29.1, this attribute only works with the `PGFPlotsX` backend.

- `lims`, `lim`, `limit`, `limits`, or `range`: This attribute with many aliases allows us to set the axis limits. We can pass `Symbol` or a tuple containing two numbers. When we use the two-element tuple, we force the boundaries of the axis to match the given numbers. The tuple can only have finite numbers; using `Inf` and `-Inf` is not allowed. Among the symbol options, we can find `:auto`, the default, that sets the axis limits following the extrema of the plotted data. Then, we can use the `:round` symbol to widen the axis by rounding the extrema values to their nearest integer. Finally, we can set `lims` to `:symmetric` to ensure that the axes are symmetric around zero – meaning that the axis extends the same distance below and above zero. When passing a two-element tuple, we can also set the `widen` attribute to `true` to widen the axis similarly to the `widen` attribute with the automatic limits.

- `widen`: This Boolean argument determines whether Plots should extend the axis limits to avoid cutting elements. By default, Plots sets it to `:auto`, extending the axes unless the user gives specific limits using the `lims` attribute.

- `link` or `links`: It determines whether to match the axis limits among different subplots of the layout. This argument doesn't have a version prefixed with x, y, or z. We can set this attribute to `:none`, `:x`, `:y`, `:both`, or `:all`. If we set this attribute to `:x`, Plots will match the limits of the *x* axes for each column of the plot layout. Similarly, choosing `:y` will connect the *y* axes for each row. To link both the *x* and *y* axes, we can set `link` to `:both`. Finally, we can set it to `:none` to avoid linking the axes or to `:all` to link all the *x* and *y* axes available in the figure, independent of their positions in the layout. We already saw an example of linking the *x* axes in the *Simple layouts* section of *Chapter 1, An Introduction to Julia for Data Visualization and Analysis*.

- `mirror`: This Boolean attribute, set to `false` by default, determines whether Plots will draw the axes on the opposite sides – the *x* axis at the top and the *y* axis on the right.

- `rotation`, `rot`, or `r`: This attribute takes a number determining the rotation angle of the tick labels in degrees. By default, Plots sets it to zero.

- `showaxis`: This attribute allows you to show or hide axes. It can take a Boolean value or a symbol. You can set it to `true`, `:yes`, `:show`, or `:all` to show all the axis – this is the default behavior. You can hide all the axes by setting `showaxis` to `false`, `:no`, `:hide`, or `:off`. Finally, you can show a specific set of axes by naming them, as x, y, and z, inside `Symbol`. For example, setting this attribute to `:x` will show only the *x* axis, while `:xy` or `:yx` will show the *x* and *y* axes.

- `ticks` or `tick`: This attribute determines the position of the ticks. You can set it using a vector of numbers, indicating the positions on the axis where Plots should place the ticks. Note that we can also use ranges, as there are subtypes of `AbstractVector`. For example, in *Chapter 10, The Anatomy of a Plot*, we set the `ticks` attribute using the `0:0.2:1` range. We can set the position of the ticks and their labels using a two-element tuple. The first element should be the `tick` values, and the second should be the `tick` labels. We can set `ticks` to `:auto` to let Plots calculate the ticks automatically for us. When using *interactive backends*, such as `Plotly` or `Plotly.js`, we can set `ticks` to `:native` to let the backend calculate the ticks. That will allow the ticks to update after zooming and panning accordingly. Finally, we can set `ticks` to `false`, `:none`, or `nothing` to turn off the ticks, or to `true` or `:all` to add them.

- `tick_direction`, `tick_dir`, `tick_or`, `tick_orientation`, `tickdir`, `tickdirection`, `tickor`, or `tickorientation`: This attribute with many aliases determines the direction of the axis ticks. By default, Plots shows the ticks inside the plotting area, as it sets this argument to `:in` by default. You can set it to `:out` to draw the ticks outside the plotting area or to `:none` to avoid showing the ticks.

These are not all the attributes available to customize plot axes; there are more. We described some of them in *Chapter 10, The Anatomy of a Plot*. In particular, we discussed there the use of `minorticks` and `minorgrid`. Plots sets both attributes to `false` to avoid showing the minor ticks and their grid. Both grids and minor grids have specific attributes to customize their aspect. Those attributes and their aliases have the `grid` and `minorgrid` prefixes respectively. Then, those attributes and their aliases end with the following strings:

- `alpha`: `gridalpha` and `minorgridalpha` determine the opacity of the grid and minor grid lines respectively. Those attributes take a number between `0.0` and `1.0`, indicating the opacity fraction, with `0.0` fully transparent and `1.0` fully opaque. By default, the Plots package sets those values to `0.1` for grids and `0.05` for minor grids. The `gridalpha` attribute has the following aliases: `gridopacity`, `gopacity`, `galpha`, `ga`, and `gα`.

- `linewidth`, `_linewidth`, `lw`, `_lw`, `width`, or `_width`: These attributes take a number determining the grid's or minor grid's line widths in pixels. Plots sets those attributes to `0.5` by default.

- `style`, `_style`, `linestyle`, `_linestyle`, `ls`, or `_ls`: This determines the style of the grid and minor grid lines. Plots set their styles to `:solid` by default, but you can change it to `:dash`, `:dot`, `:dashdot`, or `:dashdotdot`.

In this section, we have seen some attributes to customize the axes of our plots. That's particularly important, as the axis allows us to map the location of the plotted elements to the represented data values. The mapping for other data values, particularly for categorical and ordinal variables, is usually described in the figure's legend. Therefore, in the following section, we will mention some attributes that allow us to customize those legends.

Tailoring legends

The `Plots` package shows, by default, a legend with a label for each series. This section will discuss how we have to tailor those legends to match our needs. As we saw in *Chapter 10, The Anatomy of a Plot*, legends are part of Plots' subplots. Therefore, as discussed in this chapter's *Exploring plot attributes* section, we can look for the related attributes in `plotattr(:Subplot)`. All those attributes have the `legend` prefix. Note that the legend attribute names use underscores to separate the words, but all offer aliases without underscores. For example, the `legend_font_color` attribute has the `legendfontcolor` alias.

As we previously explored in this chapter's *Formatting the fonts* section, we can determine the font used for legend labels and titles using attributes. Those attributes have the following prefixes: `legendfont` or `legend_font` and `legendtitlefont` or `legend_title_font`. For example, we can use `legend_font_color` and `legend_title_font_color` to set the color of the label and title text respectively. We can also change other legend-related colors, but we will see that later in this chapter. Let's explore two practical attributes that the `Plots` package offers to customize our legends – `legend_title` and `legend_position`:

- The `legend_title` attribute has the following aliases: `legendtitle`, `legend_titles`, `legendtitles`, `leg_title`, `legtitle`, `key_title`, `keytitle`, `label_title`, and `labeltitle`. It is set to `nothing` by default; therefore, `Plots` does not show a title for our legend. We can add such a title by setting this attribute with a `String` object.

- The `legend_position` attribute is crucial when dealing with legends, as it determines their position in the plot and whether `Plots` should show them. Therefore, it has many aliases: `legendposition`, `legend_positions`, `legendpositions`, `legends`, `legend`, `leg`, and `key`.

  Setting `legend_position` to `false`, `nothing`, `:none`, or `:no` will make Plots hide the legend. Setting it to `true`, `:best`, `:both`, `:all`, or `:yes` will show the legend in the default position. By default, `Plots` sets `legend_position` to `:best`, which is `:topright` for the *GR backend*.

We can use `Symbol` objects to indicate the legend position – the supported symbols depend on the backend. We can create those symbols following the following rules:

1. First, we should decide whether we want to place the legend inside, the default, or outside the plotting area. To make the legend appears outside the plotting area, we need to include the `outer` prefix to our `Symbol`.

2. Then, we should indicate the location of the legend in the vertical direction. We can include the word `bottom` or `top` in our `Symbol` for that. The legend will be in the middle if we do not have any of those two words.

3. Finally, we indicate the position in the horizontal direction by finishing with `left` or `right`; again, not ending with those centers the legend. For example, `:outertopright`, `:outertop`, `:outerright`, `:topright`, `:top`, and `:right` are all valid options.

The following code creates all those combinations; you can execute it at the *Julia REPL*:

```julia
positions = vec([
    Symbol(join(words))
    for words in Iterators.product(
            ["", "outer"],
            ["bottom", "", "top"],
            ["left", "", "right"]
    )])
```

Here, we use the `product` iterator to generate the word combinations. This list comprehension will be a matrix that we convert into a vector using the `vec` function. To center a legend along a dimension, we must avoid including a word for that dimension. We do this by explicitly including an empty string, `" "`, in the options. We also use an empty string for the prefix to be able to plot inside the plotting area. Therefore, one iteration will `join` three empty strings and create an empty `Symbol` object, `Symbol("")`. This code also produces another forbidden combination, `:outer`. So, let's delete those from the vector of positions:

```
filter!(symbol -> !(symbol in [Symbol(""), :outer]),
    positions)
```

There are two more symbols that we can use to set the positions of the legend:

- First, we have `:inside`, which places the legend in the middle of the plotting area.
- Then, we have `:inline`, which is a little unique, since instead of constructing a legend box, it places the labels as annotations at the last point in the series.

Let's add them to our vector of positions, and also add `:none` for completeness:

```
append!(positions, [:inside, :inline, :none])
```

Now that we have the symbols to determine the legend and label positions, let's plot them by running the following code in the *Julia REPL*:

```
using Plots
plot(
    [ plot(cos, title=":$pos", legend_position=pos)
        for pos in positions ]...,
    size=(1000, 800),
    titlefontsize=12)
```

This code creates the following figure, showing the different positions we can indicate using those symbols:

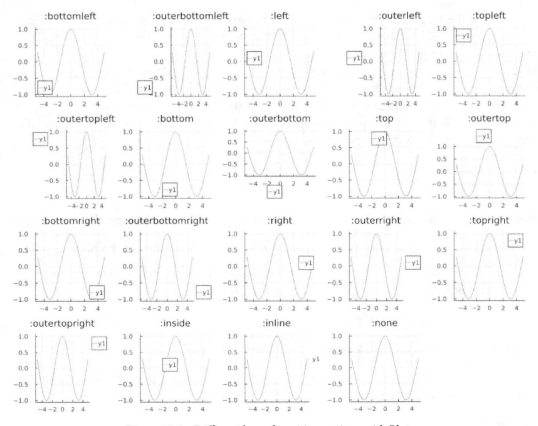

Figure 12.5 – Different legend position options with Plots

We have described how to use the legend_position attribute using Bool, Nothing, and Symbol objects, but there are other options.

We can, for example, set this attribute using a tuple containing two floating-point numbers between 0.0 and 1.0. This tuple will indicate the position of the left corner of the first legend label. The numbers are a fraction of the subplot plotting area in the horizontal and vertical dimensions respectively.

Another option is to give a number indicating an angle, in degrees, from the center of the plotting area. Plots will place the legend at that angle, inside the plotting area.

Finally, we can give a tuple containing an angle in degrees as the first element and the :inside or :outer Symbol object as the second element. That will locate the legend at that angle, inside or outside the plotting area, according to the given symbol. This way, setting legend_position to 90 or to (90, :inside) will have the same effect, placing the legend at the top.

We have discussed how to customize and place our legends using Plots. Suppose you want to know the available attribute when working with *Makie*; in that case, we recommend looking for the documentation of the Legend type. Let's now see how to deal with colors in Plots and *Makie*.

Coloring our figures

This section will teach you to select and create colors and color schemes for Plots and *Makie*. Both packages base their color utilities on the Colors and ColorSchemes packages. So, Plots and *Makie* accept any Colorant or named color that the Colors package can parse as Colorant. In particular, the most popular way to set colors in those packages is by using a Symbol or String object containing the color name. For example, :red and "red" are valid indications to select the color red. You will find the list of named colors in the Colors package documentation at https:// juliagraphics.github.io/Colors.jl/stable/namedcolors/. The Colors package supports CSS/SVG and X11 color names – it uses only the X11 names that do not clash with the CSS/SVG ones.

In the case of the Plots package, we can set a color using false or nothing to assign a fully transparent color. The Plots package also re-exports the Colorant type and the colorant string macro from Colors. Therefore, you can also choose the color red by calling parse(Colorant, "red") or colorant"red". You can give to parse or @colorant_str other strings rather than the color names. In particular, they can parse any *CSS color specification* other than currentcolor. Some of the most helpful are the colors in *hex (hexadecimal) notation* and **RGB (red, green, and blue)** values – for example, the red color can also be colorant"#FF0000" and colorant"rgb(255, 0, 0)". In the last example, we have used values between 0 and 255 for the red, green, and blue channels, but you can also use percentages – for example, colorant"rgb(100%, 0%, 0%)". To use the colorant string macro, or the Colorant type, with Makie, you need to use the Colors package explicitly.

Another helpful way to define colors is using the RGB and RGBA functions from Colors that Plots and *Makie* re-export. The RGB function uses fractions rather than percentages; therefore, it takes floating-point numbers between 0.0 and 1.0. For example, we can create the same red color in Plots and *Makie* by calling RGB(1.0, 0.0, 0.0). When using the RGB function, we can only define fully opaque colors; we need the RGBA function to use transparencies. The RGBA function has an extra positional argument, determining the value for the *alpha channel*. This channel also takes a value between 0.0 and 1.0, where 0.0 means fully transparent and 1.0 (the default if we call the function with three arguments) fully opaque. For example, RGBA(1.0, 0.0, 0.0, 0.5) will create a transparent red color that is 50% transparent/opaque.

Similarly, `Colors` can also parse color strings with alpha values – for example, we can create the same color using `colorant"rgba(255, 0, 0, 0.5)"` or `colorant"rgba(100%, 0%, 0%, 50%)"`. The alpha channel should be a fraction or percentage; for that reason, we indicate the alpha value in the first example to use a floating-point number rather than an integer between `0` and `255`, as in the other channels. In the case of *Makie*, we can also use a two-element tuple, where the first element is the color and the second element is the alpha value – for example, `(:red, 0.5)`.

Now that we have seen how to define individual colors, let's discuss the work with color palettes and gradients. `Plots` and *Makie* can use the color schemes defined in the `ColorSchemes` package. Those packages also define some color schemes of their own, but let's see the ones they have in common. The color schemes in the `ColorSchemes` package have names we can set using `Symbols`. We can see a list of them in the documentation of `Plots`, *Makie*, or `ColorSchemes`. To explore them from Julia, we will need to use the `ColorSchemes` package to access its `findcolorscheme` function. For example, we can see a list of the color schemes designed for people with **color-vision deficiency (CVD)** by running the following code:

```
import ColorSchemes
ColorSchemes.findcolorscheme("cvd", search_notes=false)
```

This will show all the color schemes in the *CVD category*. You can omit `search_notes` or set it to `true`, the default, and look for palettes including `cvd` in their notes. That will also show, for example, the `tableau_colorblind` palette, annotated as color-blind friendly. Once you import `ColorSchemes`, you can access a color scheme's notes by accessing the `notes` field of the `ColorScheme` object – for example, `ColorSchemes.tableau_colorblind.notes`. The `ColorScheme` objects also have a `colors` field containing the array of colors that define the scheme.

When mapping numerical data values to colors, especially for *scientific purposes*, it is better to choose **perceptually uniform** color schemes. Using those schemes, we can perceive equal steps in the data as equal steps in the color space. We can get a list of such color schemes by running `findcolorscheme("uniform")`. However, that list is incomplete, as some perceptually uniform color schemes don't have the `uniform` word in their notes. For example, that call will not list the color schemes from the *colorcet category*, a collection of perceptually accurate colormaps. That list also excludes the perceptually uniform color schemes from the *cmocean category*. The *scientific category* is also interesting, as their color schemes are perceptually uniform and color-blind friendly. Thankfully, we can get a list of those color schemes by running the `findcolorscheme` function with `"colorcet"`, `"cmocean"`, or `"scientific"` as input.

We will see in some color scheme notes that we can classify them into different categories. In particular, we can organize color schemes into *sequential*, such as `viridis`, or *diverging*, such as RdBu. In a diverging color scheme, we expect the extremes to be similar in lightness. Then, some color schemes can have a *categorical* and a *continuous* version. For categorical color schemes, we expect them to have perceptually distinct colors.

The `ColorSchemes` package exports a series of color schemes we can use. However, when using `Plots` and *Makie*, we can create our color schemes using the `cgrad` function. The `Plots` package also offers the `palette` function to create a color palette. The `cgrad` and `palette` functions take any of the following inputs as their first positional argument:

- A `Symbol` with the name of a color scheme from the `ColorSchemes` package – for example, `:viridis`.

- A `ColorScheme` object, such as `ColorSchemes.viridis`.

- A `ColorGradient` object, such as the ones constructed using the `cgrad` function.

- A `ColorPalette`, as constructed with the `palette` function. Note that to use the `palette` function with *Makie*, you should import the `PlotUtils` package.

- A `Vector` of colors; the colors can be `Colorant` objects or any `Symbol` or `String` that the `Colors` package can parse into `Colorant`. For example, `[colorant"red", colorant"white"]`, `[:red, :white]`, and `["rgb(255, 0, 0)", "rgb(255, 255, 255)"]` are equivalent.

The `cgrad` and `palette` functions have the `rev` and `alpha` keyword arguments. The `rev` attribute is Boolean; set it to `true` to reverse the order of the colors. The `alpha` attribute takes a floating-point value between `0.0` and `1.0`, determining the value of the alpha channel for all the colors. Both functions have more positional and keyword arguments. Let's explore them with `Plots` in a new Pluto notebook:

1. Run the following code in the first cell:

   ```
   using Plots
   ```

   We do not need to load another package, as `Plots` exports both the `cgrad` and the `palette` functions. Makie backends also export the `cgrad` function, so you can perform the steps using cgrad with Makie instead of Plots. However, using the `palette` function with Makie will require you to load the `PlotUtils` package.

2. Execute the following code in a new cell:

   ```
   cgrad(:RdBu)
   ```

We can easily see a color scheme in a Julia notebook using the `cgrad` function. This code will show us the `RdBu` color scheme from the `ColorSchemes` package.

3. Execute the following code in a new cell:

```
cgrad(:RdBu, scale=:exp)
```

We can change the scale of our colormap using the `scale` keyword argument of `cgrad`. It can take one of the following `Symbol` types: `:log`, `:log10`, `:log2`, `:ln`, `:exp`, and `:exp10`. The argument can also take `nothing`, the default, or a custom function.

4. In a new cell, execute the following code:

```
cgrad(:RdBu, 0.75)
```

The second positional argument of `cgrad` can take a number or vector of numbers. If we use floating-point numbers between `0.0` and `1.0` for a continuous color gradient, the values indicate where the color positions in the gradient. In this example, the floating-point number `0.75` indicates the position of the central color in our diverging color gradient.

5. Execute the following code in a new cell:

```
cgrad(:RdBu,
      [0.4, 0.5, 0.6, 0.7, 0.8, 0.9],
      categorical=true)
```

We set the `categorical` keyword argument to `true`. Consequently, we get `CategoricalColorGradient` rather than `ContinuousColorGradient`. By default, `cgrad` sets the `categorical` keyword argument to `nothing`, returning a continuous color map. While the vector of floating-point numbers in the second argument indicates the position of the colors for continuous color gradients, it also determines where colors start and end in categorical color gradients. The `cgrad` function automatically adds the extremes, `0.0` and `1.0`, to the vector if they aren't present. Therefore, we feed a vector with six values, but we get seven colors, and the first one goes from `0.0` to `0.4`.

6. In a new cell, execute the following code:

```
cgrad(:RdBu, 7, categorical=true)
```

Here, we create an evenly spaced categorical color map, passing an integer number to the second positional argument of `cgrad` and setting `categorical` to `true`. Each color is chosen equidistant from the color scheme.

7. In a new cell, run the following code:

```
palette(:RdBu, 7)
```

The CategoricalColorGradient created in the previous step maps a continuous variable to categorical colors. However, if we want to map categorical values to categorical colors, ColorPalette is better suited. We can use the palette function, instead of the cgrad function, to that end. The second positional argument of the palette function is the number of desired colors in our palette – 7, in this case.

At the end of those steps, our notebook will look like the screenshot in the following figure:

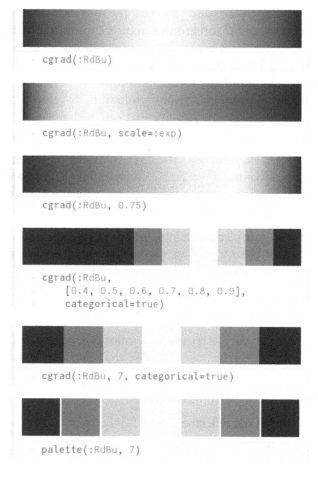

Figure 12.6 – Examples of the cgrad and palette functions

Now that we have learned to work with colors and color schemes in `Plots` and *Makie*, let's explore the `Plots` attributes to set colors.

Color attributes

`Plots` offer multiple attributes to determine the color of the different parts that compose a plot. Those attributes end with the `color` suffix or have the word `color` inside, and they follow a hierarchy. In this attribute hierarchy, the most specific attributes inherit their default values from the most general ones. To achieve this, `Plots` is set to use `:match` to match the color value of the attributes in the lower part of the hierarchy. This means that their colors match the ones of the elements that are higher in the color attribute hierarchy.

The hierarchy of color attributes is as follows. There are two main Plot attributes, `background_color` and `foreground_color`. The former is set to white by default and the latter is set to `:auto`, so the foreground color is chosen automatically based on the background color. Therefore, it is easy to have a plot where the background is black and the foreground elements are white, setting only `background_color` to `:black`. The attributes to set the background and foreground colors of a subplot have the `_color_subplot` suffix. Those subplot attributes inherit their values from the ones in the Plot object.

Then, `background_color_subplot` determines the background color inside the plotting area and inside the legend box. The last two can be modified using the attributes that have the `background_color` prefix and the `_inside` and `_legend` suffix, respectively. The background color outside the plotting area is determined by the `background_color_subplot` attribute, which inherits from `background_color`.

The `foreground_color_subplot` attribute inherits many attributes, such as the ones that have the `foreground_color` prefix. Among those, we can find the following suffixes:

- `_legend`: The `foreground_color_legend` or `legend_foreground_color` attribute determines the color of the legend border.

- `_grid` and `_minor_grid`: Determines the color of grids and minor grids, respectively. Note that the alpha value of those elements is set to a very low value by default.

- `_axis`: Determines the color of the axis ticks.

- `_text`: Sets the color of the tick labels.

- `_border`: Sets the color of the axis spines.

- _guide: Allows us to change the color of the axis labels.
- _title: Sets the color for the title font.

markerstrokecolor also matches foreground_color_subplot, allowing that element to be highlighted from the background. There are also other attributes for the series – markers, paths, and bars. Those other attributes inherit from seriescolor. In particular, they have the color suffix and the following prefixes:

- line: The linecolor attribute sets the line color for paths and bar strokes.
- fill: The color filling the paths and bar plots.
- marker: Sets the color inside the markers.

For those series attributes, there are also attributes with the same prefix and the alpha suffix to determine the value of the alpha channel for those elements. However, you can always set the alpha value together with the color, as we described previously in this chapter.

Note that seriescolor is set to :auto by default. This means that each series will take a color value from the palette determined with the color_palette attribute. This last attribute can take a vector of colors, a color gradient, or be set to :auto by default. This creates a color palette of distinguishable colors that contrast with the background.

If we want a series to be colored according to some values and a color gradient, we will need to set the desired gradient using one of the attributes containing the color suffix, such as linecolor and markercolor. Then, the values that should be mapped into colors should be given to the attributes that have the _z suffix, such as line_z and marker_z. We can see an example of this in the following code:

```
using Plots
scatter(1:10, markersize=10,
    background_color=:black,
    foreground_color_text=:grey70,
    foreground_color_legend=:grey30,
    markercolor=:viridis,
    marker_z=1:10,
    thickness_scaling=1.5)
```

In this code, we select the viridis color scheme for markercolor, and then we determine the color of each point according to the values in marker_z. In this example, we also set the background to black, making the foreground white. Then, we change the inherited white color of ticks and legend. In particular, we choose a light gray color for the ticks labels and a dark gray color for the legend box. We use the `thickness_scaling` attribute to increase line widths and font sizes to improve their visibility. The following figure shows the created figure:

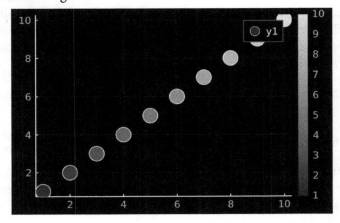

Figure 12.7 – The Plots image with custom colors

As we can see in the figure, `Plots` adds a color bar to allow mapping between the `marker_z` and the color values.

In this section, we learned the attributes that the Plots package offers to set the color of different plot elements. We can now use them to make our figures attractive, accurate, and accessible, thanks to the variety of colors and color palettes described before.

Summary

In this chapter, we have learned many attributes and utilities to better customize our figures with `Plots` and *Makie*, delving deeper into those subjects with the former. As we can use many attributes in both packages to customize our figures. Now, we can customize the aspect of fonts and colors and use LaTeX elements. We have learned to select colors and color schemes and create different color gradients. As axes and legends are crucial for mapping input data and its visual presentation, we have discussed ways to customize those elements.

In the next chapter, we will use the knowledge acquired in this chapter to work with themes to allow us to reuse our plot customizations.

Further reading

`Plots` and *Makie* offer a vast number of attributes to customize our plots; it is impossible to fit all of them in this chapter. Therefore, we strongly recommend looking into the documentation of these packages for more information:

- `Plots`: `https://docs.juliaplots.org/stable/`
- *Makie*: `https://makie.juliaplots.org/stable/`

13
Designing Plot Themes

In the previous chapter, we saw many attributes that can be useful when customizing individual plots. In this chapter, we will learn how to reuse such customizations with the help of plot themes. In particular, we will explore themes with `Plots`, *Makie*, and `Gadfly`. We will see how to find and set their predefined themes. But also, we will learn how to create reusable themes to help us produce publication-quality figures.

In this chapter, we're going to cover the following main topic:

- Working with themes

Technical requirements

For this chapter, you will need the following:

- Julia 1.6 or higher
- Pluto
- A web browser
- An internet connection

The code examples and the Pluto notebooks, including their HTML versions with embedded outputs, are in the `Chapter13` folder of the book's GitHub repository: `https://github.com/PacktPublishing/Interactive-Visualization-and-Plotting-with-Julia`.

Working with themes

In this chapter, we will work with plot themes using `Plots`, *Makie*, and `Gadfly`. These packages define themes by using the default value for each attribute. In fact, for them, a theme is a set of default attribute values we can reuse. The attributes we can determine using themes change from package to package. Still, usually, we can set colors and fonts for all of them. When looking for plot attributes to customize, you will find it helpful to look at *Chapter 10, The Anatomy of a Plot*, and *Chapter 12, Customizing Plot Attributes – Axes, Legends, and Colors*.

There are many reasons we could find it interesting to change the package's default theme; one classic example is to match the style of our IDE. Sometimes, it could be helpful to do it just to fit a personal or professional brand style. In other cases, we could find it beneficial to make our plot style closer to one from another library. Let's imagine we are using Julia to create the figures for a scientific publication, and we must meet the style required by the scientific journal. It would be great to create a theme following those requirements and use it for the different figures.

Defining and using a theme, rather than copying and pasting attributes between calls to plotting functions, follows the **don't repeat yourself (DRY)** principle. For this chapter example, we will try to create a theme that, if possible, does the following:

- Can be exported at 1200 **dots per inch (dpi)**
- Uses a sans serif font with a minimum size of 10 points
- Has a minimum line width of 0.5 pixels
- Uses a perceptually uniform color scheme

Let's start creating it with `Plots`.

Using and defining themes with Plots

Plots offers the theme function to load themes of the PlotTheme type as defined in the PlotThemes package. As the Plots package re-exports PlotThemes, we can access all the theming utilities from Plots. The PlotThemes package contains the functions and types needed to create new themes. It also offers a series of predefined themes we can use; you can see the list in the Plots documentation. Each theme has a Symbol designating its name.

Among those themes, we can find :sand, :ggplot2, and :wong, a color-blind-friendly theme. The theme function takes a Symbol, indicating the name of the theme we want to load. This function will use the default function under the hood to change the attributes' default values. So, for example, you can run theme(:wong) to activate the :wong theme. All the plots created after the theme activation will use the theme defaults. Then, to return to the Plots default theme, you can run theme(:default).

The Plots package offers the showtheme function to explore the different themes without loading them. It takes a Symbol indicating the name of the theme we want to see. For example, you can run showtheme(:wong) to witness how the :wong theme looks.

Now, let's explore how to define and use a Plots theme using a Pluto notebook:

1. Run using Plots in the first cell to load the package.

2. Create a new cell and run the following code:

```julia
journal_theme = PlotThemes.PlotTheme(
    dpi = 1200,
    tickfontsize = 10,
    legend_font_pointsize = 10,
    annotationfontsize = 10,
    guidefontsize = 12,
    legend_title_font_pointsize = 12,
    titlefontsize = 14,
    linewidth = 1,
    palette = :glasbey_category10_n256,
    colorgradient = :linear_bgy_10_95_c74_n256)
```

This creates a `PlotTheme` object defining our theme. We can pass any `Plots` attributes to the `PlotTheme` constructor as keyword arguments and use them to indicate the desired default values. Here, we want our theme to change the default `dpi` value from 100 in the default theme to `1200`. Note that any attribute we do not set here will use the default value in the `:default` theme. Therefore, we do not set `fontfamily` to `"sans-serif"` as that is the `Plots` default.

Then, we change the default font sizes so that the minimum is 10 points. After that, as of `Plots` version 1.31.1, there is no way to set the line width of all the plot elements, so we set line widths for the series. We can also set `gridlinewidth` and `minorgridlinewidth`, but they are `0.5` pixels by default. We selected color schemes for our categorical palette and color gradients using the `palette` and `colorgradient` keyword arguments, respectively. As you will note, `colorgradient` is not a `Plots` attribute; it has a special meaning in the theme context defining the default color gradient. Here, we have chosen two color schemes from the **colorcet category**, which are perceptually uniform. You can read more about color schemes and font attributes in *Chapter 12, Customizing Plot Attributes – Axes, Legends, and Colors*.

3. Run the following code in a new cell:

```
add_theme(:journal_theme, journal_theme)
```

We added our theme to `Plots` by running the `add_theme` function. This function takes a `Symbol` defining the theme name and the `PlotTheme` object containing the theme specification. In this case, we named our theme `:journal_theme`. Now that we have registered our theme locally, we can see it and use it.

4. In a new cell, run the following code:

```
showtheme(:journal_theme)
```

It creates an example plot showcasing our new theme, as shown in the following figure:

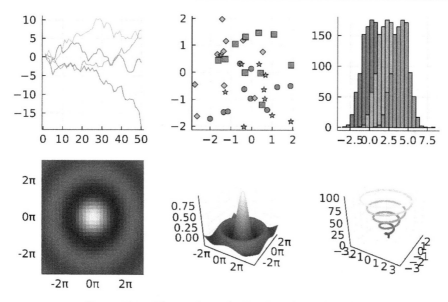

Figure 13.1 – Theme shown by the showtheme function

5. Run theme (:journal_theme) in a new cell to activate our theme; after that, all the newly created plots will use the new theme.

6. Execute the following code in a new cell to create a plot:

```
plot([sin cos], labels=["sin" "cos"],
    title="Title", xlab="x", ylab="y",
    linewidth=2)
```

It uses the new theme to create the plot shown in *Figure 13.2*. Note that we have changed the default line width of the plot, set to 1 pixel by the theme, to 2 pixels using the linewidth attribute. As plot themes only set default values, we can always overwrite the attribute values in the call to the plotting function.

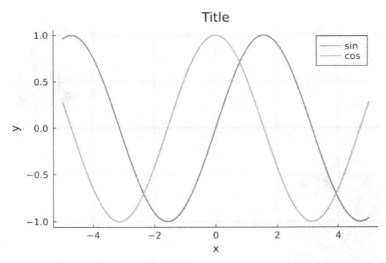

Figure 13.2 – Plots figure using a custom theme

We have seen how to create and use a theme with `Plots` in the previous steps. There are some interesting attributes and functions that Plots offers that we will find handy when saving figures for publications, posters, or slides. The first is the `thickness_scaling` attribute, which scales the plot's line widths and font sizes. If you only want to scale the size of the fonts, you can call the `scalefontsizes` function. After executing that function, you can return to the previous behavior by calling the `resetfontsizes` function.

Now we know how to work with themes using `Plots`, let's explore how to do the same using *Makie*.

Theming with Makie

Makie exports the `Theme` type for building new themes. Its constructor can take any plot attribute and set its default using keyword arguments. Please consider that we need to pass the attributes for a particular plot object as a *named tuple* to the keyword argument that has the type name – in *CamelCase*. For example, to set the default font size for the plot title to 18 points, we need to do `Theme(Axis=(titlesize=18,))` as the `titlesize` attribute is part of the `Axis` object. You can check the available attributes in the type documentation. The trailing comma after `titlesize` is needed to create the named tuple as we have a single attribute.

To set a particular theme, we need to use the `set_theme!` function. This function takes a `Theme` object. Optionally, we can specify new default attribute values by setting them through the `set_theme!` keyword arguments. *Makie* exports a series of predefined themes, which we can load using the functions that have the `theme_` prefix; for example, theme_minimal. As this function returns a Theme object, we can call set_theme! with the output we want; for instance, set_theme!(theme_minimal()). We can call set_theme! without arguments to return to the *Makie* default theme.

If we want to use a theme for a single plot, rather than activating and deactivating the theme using the set_theme! function, we can use the with_theme function. with_theme is a higher-order function that calls a plotting function with a specified theme. We usually create the plotting function as an anonymous function using the do syntax as explained in the *Anonymous functions* section of *Chapter 1, An Introduction to Julia for Data Visualization and Analysis*. As with the set_theme! function, with_theme also allows overriding default attribute values using keyword arguments.

Let's see an example of creating and using a theme with *Makie* using Pluto:

1. Execute the following code in the first cell:

    ```
    begin
        using CairoMakie
        import ColorSchemes
    end
    ```

    Here, we loaded *Makie* using the CairoMakie backend. We also imported the ColorSchemes package to get the color vector from its color schemes.

2. Create a new cell and execute the following code:

    ```
    journal_theme = Theme(
        resolution = (600, 400),
        fontsize = 10,
        Axis = (xlabelsize = 12,
                ylabelsize = 12, titlesize=14),
        Legend = (titlesize = 12,),
        linewidth = 1,
        palette = (color =
        ColorSchemes.glasbey_category10_n256.colors,),
        colormap =:linear_bgy_10_95_c74_n256)
    ```

    Here, we created a theme for our plot using the Theme constructor. Here, we set the default resolution to (600, 400). *Makie* doesn't offer a dpi attribute; however, you can control the dpi of a bitmap figure and the actual font sizes of a vector image using the save function.

Then, we set the default fontsize to 10 points. As this will make all the fonts have 10 points, we set larger sizes for some fonts. In particular, we made the axis labels and the plot title have 12 and 14 points, respectively. We also made bigger legend titles. After that, we set the default linewidth for the series to 1. *Makie* offers attributes such as spinewidth and xgridwidth to change the width of the Axis line elements; however, they are all set to 1.0 by default.

Finally, we set the categorical palette and the default color gradient using the palette and colormap attributes, respectively. Note that the palette attribute is special, as *Makie* looks there for cycled attributes, usually color and marker. The color attribute should take a vector of colors, so we access that vector from the colors field of the ColorSchemes package. You can read more on color schemes in *Chapter 12, Customizing Plot Attributes – Axes, Legends, and Colors*.

3. Execute the following code in a new cell:

```
with_theme(journal_theme; linewidth=2) do
    xs = collect(-5.0:0.1:5.0)
    fig, axs, plt = lines(xs, sin.(xs),
        label="sin",
        axis=(xlabel = "x",
            ylabel = "y", title = "Title"))
    lines!(axs, xs, cos.(xs), label="cos")
    axislegend()
    fig
end
```

We used the with_theme function to create a plot with our newly designed theme, taking advantage of its keyword argument to increase the linewidth attribute of the theme. We used the do syntax to create an anonymous plotting function. The following figure shows the created plot:

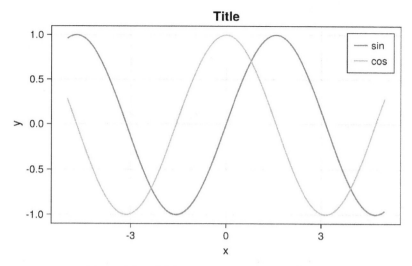

Figure 13.3 – Makie plot using a custom theme

Here, we have explored some aspects of the powerful theming capabilities of *Makie*. In the following section, we will quickly explore the use of themes with `Gadfly`.

Using themes with Gadfly

The `Gadfly` package also exports a `Theme` type that takes a series of keyword arguments in its constructor, allowing the modification of its default values. We can observe the list of attributes that we can set in the documentation for the `Theme` type. Then, `Gadfly` themes are organized in a stack, where the last theme in the stack is the active one. Therefore, we can set a theme using the `Gadfly.push_theme` function and return to the previously active theme using the `Gadfly.pop_theme` function. Like *Makie*, `Gadfly` also offers a `with_theme` function: `Gadfly.with_theme`.

A difference between `Gadfly`, `Plots`, and *Makie*, is that `Gadfly` functions don't take keyword arguments to override theme attributes. Instead, `Gadfly` offers the `style` function, which takes the same keyword arguments as the `Theme` constructor. This function overrides the attributes in the active theme and returns a `Theme` object; therefore, we can use it with the previous functions. Another noteworthy difference is that we can use `Theme` objects in `plot` calls to set their theme. We can name themes defining a new method for the `Gadfly.get_theme` function that returns the desired `Theme` object. Then, we can call the `Gadfly.with_theme` or `Gadfly.push_theme` function with a `Symbol` indicating the name of the theme. `Gadfly` comes with two predefined named themes: `:default` and `:dark`.

Summary

In this chapter, we have learned how to create themes for our plots to have reusable attribute customizations. As always, we have focused on the `Plots` package. We have also seen the basics of creating themes with *Makie* and `Gadfly`. We usually use plot themes to define fonts, colors, sizes, and other esthetical aspects. While `Plots` and *Makie* allow setting almost any attribute through their theming system, we can get even more flexibility with their plot recipe systems.

In the next chapter, we will look at the basics of creating plotting recipes with those packages.

Further reading

`Plots`, *Makie*, and `Gadfly`, offer many attributes we can access from their theming systems to customize plots. There are aspects of the `Makie` and *Gadfly* theming capabilities that we couldn't cover in this chapter. Therefore, we recommend taking a look at their documentation pages:

- `Plots`: `https://docs.juliaplots.org/stable/`
- *Makie*: `https://makie.juliaplots.org/stable/`
- `Gadfly`: `http://gadflyjl.org/stable/man/themes/`

14
Designing Your Own Plots – Plot Recipes

So far in this book, you've learned how to plot our data, customize the plot aspect, and even combine different plots to create more complex figures. In this chapter, we will use that knowledge to develop new plot types using the recipe system. The Julia ecosystem heavily uses this system, and we have already seen examples of these `Plots` recipes in this book, such as the recipes for plotting statistical plots, graphs, and maps. This last example is interesting as it also uses the shape series to draw polygons, which is helpful when drawing with `Plots` and creating custom markers — two topics we will address in this chapter.

In this chapter, we will learn how to create custom shapes and plot recipes with `Plots`. Also, we will briefly describe how to draw and create plotting recipes with *Makie*. We will introduce you to these topics so that you can create custom visualizations for your favorite data types with minimum overhead. What's more, you will be able to create entirely new plot types with the tools described in this chapter.

In this chapter, we're going to cover the following main topics:

- Drawing shapes
- Creating plotting recipes
- Writing plotting recipes for Makie

Technical requirements

For this chapter, you will need the following:

- Julia 1.6 or higher

- Pluto

- A web browser

- An internet connection

The code examples and the Pluto notebooks, including their HTML versions with embedded outputs, can be found in the `Chapter14` folder of this book's GitHub repository: `https://github.com/PacktPublishing/Interactive-Visualization-and-Plotting-with-Julia`.

Drawing shapes

The `Plots` package exports the `Shape` type to construct *polygons*. This section will teach us how to use it to create custom shapes with `Plots`. This will be helpful when drawing or while creating novel plotting types. We can also take advantage of `Shape` to design *custom markers*. The `Shape` constructor takes the vertices coordinates of the desired polygon. We can pass them in two different ways:

- We can give a vector for each axis containing the vertex coordinates for that dimension. For example, we can call `Shape(x, y)` if the x and y variables are vectors of numbers of the same length. This is one of the ways we usually call the `plot` function; that is, `plot(x, y)`.

- We can give a single vector containing a tuple of numbers, with each tuple containing a vertex's (*x, y*) coordinates. Interestingly, we can also call `plot` or other plotting functions such as `scatter` with that input. For example, `plot([1, 2, 3], [6, 7, 5])` and `plot([(1, 6), (2, 7), (3, 5)])` are going to generate the plot shown in the following screenshot:

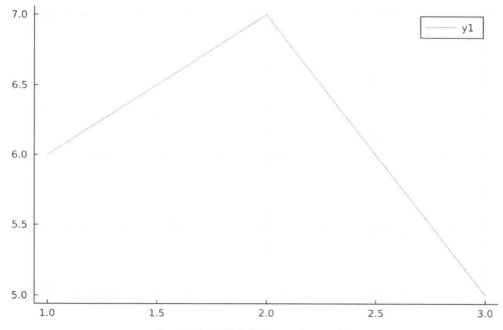

Figure 14.1 – Path between three points

The path in the preceding screenshot has three points and two segments between them. We can construct a polygon with those points as vertices by adding a segment between the last and the first point. However, it would be better to use the Shape type than add an extra point to our path to close the polygon. The plot function knows how to plot those Shape objects. It also offers the fillcolor attribute to set the color of the area inside the created polygon. For example, calling plot(Shape([1, 2, 3], [6, 7, 5])) or plot(Shape([(1, 6), (2, 7), (3, 5)])) will create the polygon shown in the following screenshot:

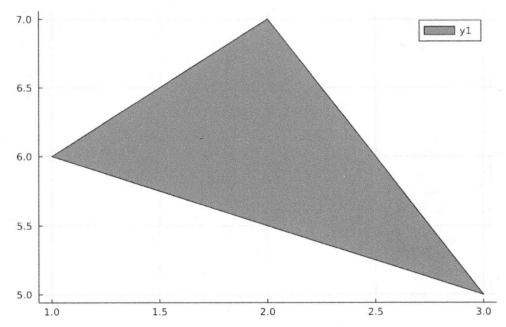

Figure 14.2 – Polygon created using the Shape type

We can draw different shapes in the same plot in two different ways. The first way is to plot a vector of Shape objects, as shown in the following code:

```
plot([
     Shape([(1, 6), (2, 7), (3, 5)]),
     Shape([(0, 0), (0, 4), (2, 4), (2, 0)])
])
```

This code creates two polygons; a triangle and a rectangle. It generates the plot shown in the following screenshot:

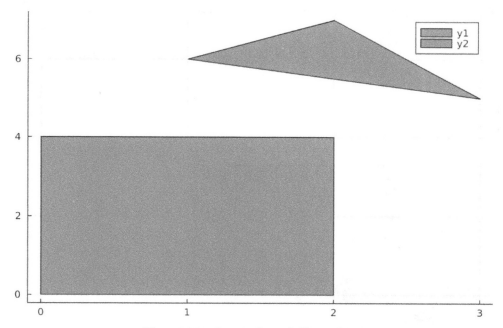

Figure 14.3 – A series for each Shape object

As we can see, each Shape object belongs to a different series, with its own color and legend label. If we want to draw all the Shape objects in a single series, we need to pass a single Shape object, where the different shapes are separated using NaN or infinite values, such as Inf. That is also how we can separate segments of a path when using the plot function – we did that to create different arrow segments in *Chapter 10, The Anatomy of a Plot*. For example, we can create a single series containing the same polygons shown in the preceding screenshot by running the following code:

```
plot(Shape([(1, 6), (2, 7), (3, 5),
     (NaN, NaN),
     (0, 0), (0, 4), (2, 4), (2, 0)]))
```

Here, we separated the vertex (x, y) coordinates of the two polygons adding a (NaN, NaN) vertex between them. Alternatively, we can use two vectors to create the same plot. The first vector contains the x axis coordinates and the second the y axis coordinates of the vertices; again, we use NaN values to separate the shapes:

```
plot(Shape([1, 2, 3, NaN, 0, 0, 2, 2],
     [6, 7, 5, NaN, 0, 4, 4, 0]))
```

Any of the two syntaxes will produce the plot shown in the following screenshot:

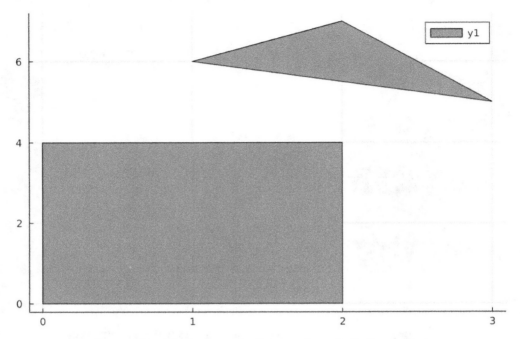

Figure 14.4 – NaN-separated polygons using a single Shape object

Another interesting aspect of Shape objects is that we can use them to create *custom marker shapes*. The chosen polygon should have (0, 0) as its *center*, and its size should be similar to the *unit circle*. Let's create a marker that looks like a dinosaur using Pluto:

1. Execute using Plots in the first cell.

2. In a new cell, execute the following code:

```
dinosaur = Shape([(-1.47,.25),(-.18, .47),(.07,.45),
(.27,.48),(.62,.43),(.92,.54),(1.17,.55),(1.27,.5),
(1.42,.5),(1.47,.38),(1.27,.38),(1.24,.3),(1.4,.2),
(1.38,.15),(1.02,.2),(.53,.04),(.54,-.05),(.5,0),
(.47,-.04),(.43,0),(.27,-.1),(.12,-.4),(.17,-.45),
(.32,-.5),(.04,-.5),(.05,-.45),(0,-.25),(-.42,-.3),
(-.48,-.4),(-.4,-.5),(-.5,-.5),(-.58,-.35),(-.5,-.2),
(-.18,-.1),(-.38, 0),(-.48,.1),(-1.47,.25)])
```

This creates a Shape object using a vector of vertex coordinates for a simple dinosaur shape center at *(0, 0)*. As of Plots *v1.31.1*, we need to close the path when creating custom markers so that the last vertex is identical to the first one. We'll look at the designed polygon in the next step.

3. Execute the following code in a new cell:

```
plot(dinosaur, aspect_ratio = :equal)
```

This plots our Shape object. Here, we set the aspect_ratio attribute to :equal to see the correct proportions of our polygon. Now, let's use our dinosaur as a marker shape for scatter plots.

4. In a new cell, execute the following code:

```
scatter(1:10, markershape = dinosaur, markersize = 35)
```

The preceding code creates a scatter plot using our polygon to determine the marker shape using the markershape attribute. In this example, we have made markers bigger than the default to appreciate the silhouettes. The following screenshot shows the plot that was created:

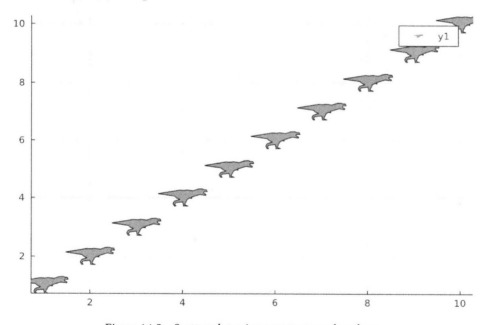

Figure 14.5 – Scatter plot using a custom marker shape

In the previous examples, we have always created shapes with line segments. However, we can also make curved objects by increasing the vertex number. We did this in the *Creating a basic plot* section of *Chapter 1, An Introduction to Julia for Data Visualization and Analysis*, when plotting curves and drawing a unit circle from a set of points.

Also, the `Plots` package exports the `BezierCurve` type to help us draw curves. You need to give the control points to the `BezierCurve` constructor. Those points should be of the `Plots.P2` type, which is an alias for the `GeometryBasics.Point2{Float64}` type; for example, `Plots.P2(0,0)`. The first point of the vector determines the start point, while the last is the curve's end point. The additional points are the other control points of the curve; you need at least one to avoid getting a straight segment. Then, you can use the `coords` function to take a given number of points from the Bézier curve and feed them to the `Shape` constructor. For example, the following code creates a `Shape` object using the `BezierCurve` type:

```
begin
    curve = BezierCurve(
        Plots.P2[(0, 0), (0.5, 1), (1, 0)])
    plot(Shape(coords(curve, 100)))
end
```

Here, we indicated that the created vector should contain objects of the `Plots.P2` type, to let Julia automatically convert our tuples. Here, we provided three control points; the start at `(0, 0)`, the end at `(1, 0)`, and an extra control point at `(0.5, 1)` – note that internal control points control the curvature and usually lie outside the curve. Then, we used the `coords` function to pass `100` points to the `Shape` constructor for plotting. Et voilà, we have the `Shape` object shown in the following screenshot, which contains a Bézier curve:

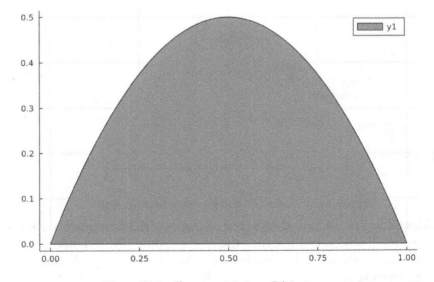

Figure 14.6 – Shape containing a Bézier curve

In this section, we have seen ways to draw polygonal shapes with `Plots`. In *Makie*, we can create polygonal shapes using the `poly` function. This function takes different inputs, from a vector of vertices, usually of the `Point2f` type, to make shapes constructed using helper types from the `GeometryBasics` package. We can find `Polygon`, `Rect`, `Circle`, `Cylinder`, and `Mesh` among those types. There are two advantages of *Makie* regarding plotting custom shapes: you can draw hollow and three-dimensional objects. Finally, note that `Gadfly` also allows us to draw polygonal shapes by taking advantage of the polygon geometry: `Geom.polygon`.

Now that we have learned how to draw custom shapes in `Plots`, let's use them to create our plot types with the plotting recipe system.

Creating plotting recipes

The `Plots` ecosystem offers the `RecipesBase` package so that you can add plotting recipes. It is handy as a package developer since it offers minimum loading overhead for your package. It has no dependencies and enables access to Plots' plotting capabilities. To that end, the package exports the `@recipe` macro, which takes a function definition that describes our recipe. We have different recipe types, depending on the type signature of the defined method, as the system uses Julia's multiple dispatch.

The `@recipe` macro has a special syntax for setting attribute values. We can use the `-->` operator to select a different default value for the attribute. This states that the attribute should take the given value if the user has not set it. We can use the `:=` operator instead to force a given value for an attribute. Both operators take the attribute on the left-hand side and the desired value on the right-hand side. For example, the following code fragment shows how those operators will look inside a `@recipe` block:

```
legend_position --> false
framestyle := :none
```

Suppose we need to use a complex expression to calculate the value we want to assign. In that case, we can wrap that expression between parenthesis. Note that we need to use the real attribute names rather than their aliases when defining a recipe. To find out those names, you can use the `plotattr` function, which we looked at in *Chapter 12, Customizing Plot Attributes – Axes, Legends, and Colors.*

Finally, we can optionally add flags after a comma on the rightmost side of the expression by setting the attribute with the `-->` operator. These flags are `:quiet`, `:required`, and `:force`. The `:quiet` flag will suppress the unsupported attribute warnings. We will find this helpful when our recipe sets an attribute that only a few plotting backends support – for example, when we set the `hover` attribute to display custom labels on mouse hover when plotting with Plotly-based backends. The `:required` flag will do the opposite, throwing an error if the backend doesn't support the attribute. Finally, the `:force` flag will set an attribute to a value that the user will not be able to change.

Inside a recipe, we can access its attributes using the `plotattributes` dictionary. Note that we can add new attributes by adding keyword arguments to our function definition. To avoid problems because of unknown attributes, you can delete the added attributes from the `plotattributes` dictionary before the end of the processing pipeline; for example, by calling the `delete!` function. What's more, inside a `@recipe` call, we can use the `@series` macro, which creates a new series by taking a copy of the `plotattributes` dictionary at that point. The `@series` macro uses the same syntax for setting attributes and returns the variables needed for plotting.

As we mentioned previously, there are four types of recipes in `Plots`. These recipes are called in a specific order in the plotting pipeline. Each of them should return the plot arguments – the positional ones – needed further in the pipeline. In the following subsections, we will see those plotting recipes in the order in which `Plots` applies them in the processing pipeline.

User recipes

The `Plots` package calls *user recipes* early in the pipeline. It allows you to process custom types or specific types combinations. The signature for this recipe is just a series of objects. For example, imagine that we have defined the following type to store the minimum and maximum values for a series of experiments:

```
struct Extrema
    min_values::Vector{Float64}
    max_values::Vector{Float64}
end
```

We can load the `RecipesBase` package and define the following user recipe for our type:

```
@recipe function f(values::Extrema)
    xs = Float64[]
    ys = Float64[]
    for (i, (min_val, max_val)) in enumerate(
            zip(values.min_values, values.max_values))
        append!(xs, [i, i, NaN])
        append!(ys, [min_val, max_val, NaN])
    end
```

```
        xs, ys
end
```

Here, we took an object instance of our type and worked with it to create the desired output for plotting. Plots discards the function's name – f, in this example – which we need for syntax purposes. Here, in the function body, we converted our object into two vectors for storing the x and y coordinates to create a series of segments. In those vectors, we separated each segment using a NaN value. Here, we used the enumerate function described in the *Exploring GraphRecipes* section of *Chapter 7, Visualizing Graphs*, to get the experiment number for the X-axis. Moreover, the graphplot function described in that chapter is a user recipe. To iterate each pair of values simultaneously, we used the zip function.

To create a plot using that recipe, we need to load the Plots package and call the plot function on an object of the Extrema type, like so:

```
using Plots
plot(Extrema(rand(100) .- 1, rand(100) .+ 1))
```

This plot call creates a plot similar to the one shown in the following screenshot:

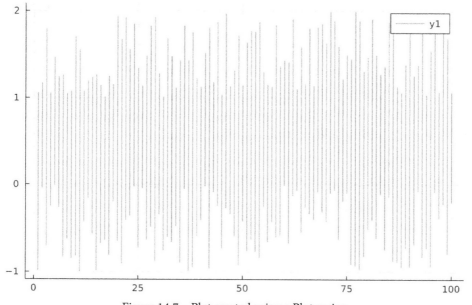

Figure 14.7 – Plot created using a Plot recipe

User recipes are valuable when creating custom visualizations. If we do not have a custom type, we can create one, but only for dispatching purposes, using the @userplot macro. This macro creates a type and exports some convenient functions, known as *shorthands*, to call its recipe. For example, executing @userplot Blackboard will create the blackboard and blackboard! functions, which call plot with an instance of the Blackboard type. Then, we need to define the recipe for that type, like so:

```
@recipe function f(object::Blackboard)
    background_color := :black
    foreground_color := :white
    palette --> palette([:gray90, :gray30] , 6)
    framestyle --> :none, :quiet
    linewidth --> 2, :required
    legend_position --> :none, :force
    object.args
end
```

As we can see, the user recipe should take the type defined by the @userplot macro. The type wraps the arguments given to the plot function in the args field. User recipes should return the arguments for the following recipe on the pipeline: the type recipes. In this example, the last line of the function body returns the content of the args field to pass the plot call arguments untouched. To only showcase the syntax to set attributes inside recipes, we used the := and --> operators. With the --> operator, we have also used the three available flags.

Once we have defined the recipe, we can call the functions created by the @userplot macro – after loading Plots. For example, executing blackboard(rand(10, 4)) will create a plot similar to the following:

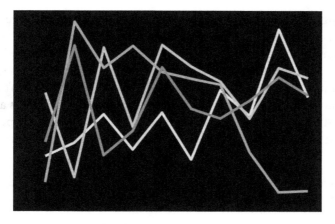

Figure 14.8 – Example of a recipe created with the @userplot macro

To see the differences in the different ways we have to set recipe's attributes, let's call the `blackboard` function again, but change the default attribute values:

```
blackboard(rand(10, 4),
    background_color=:white,
    palette=:greens,
    framestyle=:box,
    linewidth=10,
    legend_position=:outerleft)
```

That code creates the plot we see in the following figure:

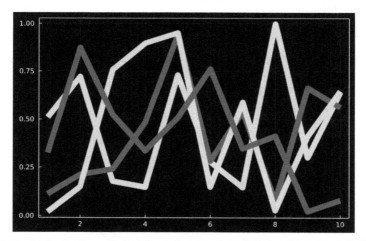

Figure 14.9 - Recipe called with attributes values differing from the defaults

As we can see, changing the background color doesn't have any effect as we have set `background_color` using the `:=` operator in the recipe. We changed the `palette` from gray to green hues, the `framestyle` from `:none` to `:box`, and the linewidth from 2 to 10, getting thicker lines. That was possible as we have set the three attributes using the `-->` operator. Note that the `:quiet` and `:required` flags don't change this behaviour — they are essential to deal with the different plotting backends as not all support the same attributes. However, we could not change the `legend_position` from `:none` to `:outerleft` because we have used the `:force` flag when setting that attribute.

Now that we have seen some examples of the `@recipe` macro syntax when creating user recipes, let's look at the other recipe types.

Type recipes

Julia makes it easy to create new types, but not all new types can be plotted out of the box. The `plot` function will try to convert our types into something it can plot. When that's impossible, the `plot` function gives an error indicating that `Plots` cannot convert our type into series data for plotting. To fix that issue, `Plots` allows us to write type recipes that define a conversion method for our type into something `Plots` can handle. For example, let's say we define a custom type that wraps a vector:

```
struct VectorWrapper
    vec::Vector{Float64}
end
```

We can create the following type recipe to indicate to the `plot` function that it should plot the wrapped vector:

```
@recipe f(::Type{VectorWrapper}, w::VectorWrapper) = w.vec
```

As we can see, the signature of a type recipe has the chosen type in the first argument and an object of that type in the second.

Plot recipes

The processing pipeline executes the plot recipes after processing the input data but before performing the plot. Therefore, we can assume that our plot recipe takes standard types that `Plots` can represent as arguments. It allows us to define `Plot` and `Subplot` attributes and add subplots. For example, we can use it to create recipes that display multiple subplots, such as marginal histograms. The signature for the function will be `f(::Type{Val{:name}}, plt::AbstractPlot)`, where we can replace `:name` for the name we want for our recipe.

Let's look at an example of a plot recipe using a Pluto notebook. In this example, we will create something similar to a rug plot. In a **rug plot**, we draw a short vertical line next to the axis for each data point. Those lines give us an idea of the distribution for the variable in that axis. A classical rug plot locates the lines inside the plotting area. However, in this example, we will draw them outside to show an example of how a plot recipe can create custom layouts. Also, to make the example short, we will only draw lines for the x-axis. Open a new Pluto notebook to execute the following steps:

1. Execute `using RecipesBase` in the first cell to access the `Plots` recipe system.

2. Create a new cell and run the following code:

```julia
@recipe function f(::Type{Val{:xrug}},
        plt::AbstractPlot; main=:scatter)
    layout := @layout[ a{0.95h}; b ]
    link := :x
    @series begin
        subplot := 1
        seriestype := main
    end
    @series begin
        subplot := 2
        seriestype := :vline
        framestyle := :none
        legend_position := false
        y := plotattributes[:x]
    end
end
```

This preceding code created a plot recipe named `xrug` using the `@recipe` macro. This recipe sets two plot-level attributes: `layout` and `link`. The figure layout will contain two plots; the main plot on top, taking 95% of the figure's height, and the rug on the bottom. We linked the limits of both x-axes to place the vertical lines correctly.

Then, we used two @series macros, each of them to fill one of the subplots. Inside those macros, we used the subplot attribute to designate where to place each plot. As each subplot will inherit the xrug series type, we set a different series type for each plot. To control the value of the seriestype attribute of the main plot, we defined the main attribute by adding a keyword argument in the recipe definition. Another interesting detail in this recipe is that we take the X-axis values from the plotattributes dictionary to set the positions of the lines. The rest of the recipe uses the := operator to force some attributes to make the plot clean.

3. Execute using Plots in a new cell to access the plot function and use the defined recipe.

4. Run the following code in a new cell to use the recipe:

```
plot(exp.(0:0.1:5), 0:0.1:5, seriestype = :xrug)
```

Here, we needed to set the seriestype attribute to the recipe's name, xrug, as a Symbol. The following screenshot shows the plot that was generated:

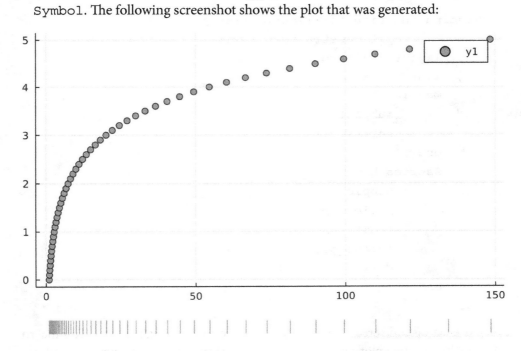

Figure 14.10 – Plot recipe example of xrug creating a plot with two subplots

5. Create a new cell and run the following code:

```
@shorthands xrug
```

The `@shorthads` macro is defining the `xrug` and `xrug!` functions in this example to ease calling this recipe.

6. Run the following code in a new cell:

```
xrug(exp.(0:0.1:5), 0:0.1:5)
```

Here, we used the newly defined shorthand to call our recipe.

In the next section, we will learn how to create recipes with *Makie*, and we will create a classical *rug plot* as an example. But, before that, let's look at the last plotting recipe in the `Plots` pipeline: the *series recipe*.

Series recipes

A *series recipe* is the last recipe that's called in the pipeline; `Plots` only calls it if the plotting backed doesn't have native support for that recipe. Therefore, they are used to extend backend support, but more generally, they can be used to create new series types. The function signature to define them is `f(::Type{Val{:name}}, x, y, z)`, where x, y, and z are mandatory to get the plotting arguments and interpret this recipe as a series recipe rather than a plot recipe.

In this section, we have briefly explored the recipe system of `Plots`. Now, we should be able to understand the syntax of the plotting recipes out there. You should also be able to create custom visualizations for your types by using `RecipesBase` from the `Plots` ecosystem.

In the next section, we will see that *Makie* also offers a recipe system. As you will see, the recipe system of *Makie* has many similarities to the one described in this section for `Plots`.

Writing plotting recipes for Makie

Makie also offers a lightweight recipe system thanks to the `MakieCore` package. This package only depends on the `Observables` package to allow interactivity. As of `MakieCore` *version 0.3.5*, *Makie* provides two kinds of recipes: type and full recipes. **Type recipes** in *Makie* are similar to the ones of `Plots`, as they can convert a user-defined type into something *Makie* can plot. They also allow us to specify the default plot type for that custom type. **Full recipes** in *Makie*, on the other hand, are like Plots' series recipes. *Makie* uses this full recipe to define many plot types, such as `hlines` and `vlines`. Sadly, they do not offer a high-level API to control the plot layout as Plots' plot recipes do.

Type recipes are defined thanks to the `convert_arguments` function. When *Makie* calls the `plot` function with an unknown type, it will try to convert it into something it can plot by calling that function using a `PlotType`; for example, `Scatter`, in the first argument. If Julia does not find a method with that signature, it will call `convert_arguments` using a `ConversionTrait`, such as `PointBased`, which includes plot types such as `Scatter` and `Lines`. Let's look at an example of a *Makie* type recipe while reproducing the `Plots` user recipe example:

1. Create a new Pluto notebook and execute the following code in the first cell:

    ```
    using MakieCore
    ```

    The preceding code loads the `MakieCore` package, which is the only one we need for defining *Makie* recipes.

2. In a new cell, execute `using GLMakie` to load the backend we will use to create the plots.

3. Execute the following code in a new cell:

    ```
    struct Extrema
        min_values::Vector{Float64}
        max_values::Vector{Float64}
    end
    ```

    The preceding code creates the type we will use as an example – the same one we used in the *User recipes* section of this chapter.

4. In a new cell, execute the following code:

    ```
    MakieCore.plottype(::Extrema) = LineSegments
    ```

    Here, we created a method for the `plottype` function that takes an instance of our type as input and returns the plot type we want to use for our type. In this case, we want to use the `LineSegments` plot type to create a plot similar to the one shown in *Figure 14.7*.

5. Execute the following code in a new cell:

```
function MakieCore.convert_arguments(
        ::Type{<:LineSegments},
        values::Extrema)
    points = Point2f[]
    for (i, (min_val, max_val)) in enumerate(
        zip(values.min_values, values.max_values))
        push!(points, (i, min_val))
        push!(points, (i, max_val))
    end
    (points,)
end
```

Here, we defined a method for the `convert_arguments` function from `MakieCore` that takes a `LineSegments` plot type and an instance of our type. The function should return a tuple containing objects that the related plotting function can take – note the tailing coma in the returned object to ensure it is a tuple.

In this case, `linesegments` can take a vector of points with an even number of elements – odd elements are the segments' start points, while even elements are the endpoints. Since we defined the points vector as containing elements of the `Point2f` type, each tuple is converted into this type when we push it into the points vector.

Note that the code is similar to the one shown in the `Plots` example. The main difference is that *Makie* doesn't need to use `NaN` values to separate the segments when using the `linesegments` function, making the code more straightforward.

6. In a new cell, run the following code:

```
plot(Extrema(rand(100) .- 1, rand(100) .+ 1))
```

The preceding code calls the type recipe we created in the previous steps using random values. The following screenshot shows the plot that was generated with this call:

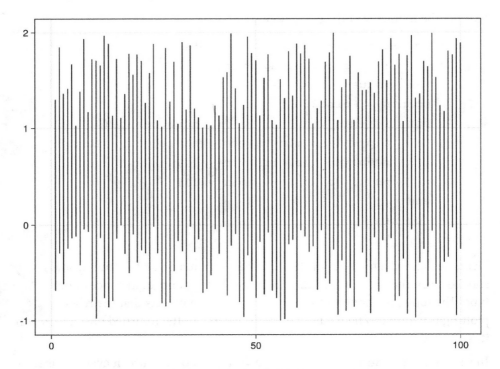

Figure 14.11 – Example of a Makie type recipe

Now that we have learned how to create type recipes with *Makie*, let's learn how to create a *Makie full recipe* by creating a classical *rug plot*:

1. Open the same Pluto notebook we used to develop our type recipe; we need the same packages.

2. Create a new cell and execute the following code:

```
begin
    @MakieCore.recipe(Rug, x, y) do scene
        Attributes()
    end

    function MakieCore.plot!(plot::Rug)
        vlines!(plot, plot[1], ymax=0.025)
        hlines!(plot, plot[2], xmax=0.025)
    end
end
```

The preceding code creates our full recipe. First, it uses the @recipe macro from MakieCore to generate a plot type – Rug, in this example. Inside @recipe, we can use Attributes to define a set of default attribute values for our recipe and create recipe-specific attributes. The macro also defines two functions, using the plot type in lowercase. In this case, the functions are rug and rug!, where the latter can plot over a pre-existing axis.

Then, we defined a new method for plot! that takes a plot of our type. Inside the function body, we can use any plotting function from *Makie*. In this example, we used hline! and vline! to draw the lines of the rug plot. We can access the input argument by indexing the plot object using the argument position. For example, plot[1] gives the coordinates on the *X*-axis while plot[2] gives us the *Y*-axis coordinates. An interesting aspect is that both values are Observable objects, allowing our recipe to be interactive. Also, the Attributes object converts any default value into an Observable. Let's use our rug plot recipe and test its interactivity.

3. Execute the following code in a new cell:

```
begin
    x = Observable(rand(100))
    y = Observable(rand(100))
end
```

The preceding code defines two vectors of random numbers for this example. We define them as Observable objects to allow GLMakie to interactively redraw the modified pieces of our plot.

4. In a new cell, run the following code:

```
begin
    fig = Figure()
    ax = Axis(fig[1, 1])
    scatter!(ax, x, y)
    rug!(ax, x, y)
    display(fig)
end
```

The preceding code creates our rug plot. Here, we chose to use a scatter plot as our main plot, and we used the `rug!` function to add the lines that show the distribution of the two variables next to the axis spines. We called the `display` function at the end to show our plot in an interactive *GLFW window*. The following screenshot shows the rug plot that was created in this full recipe:

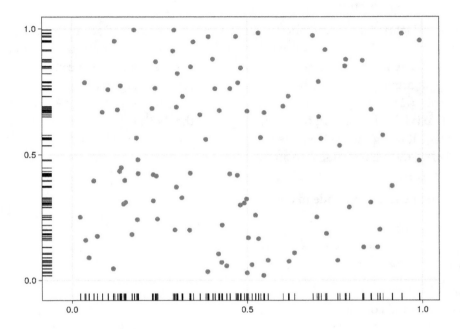

Figure 14.12 – Rug plot recipe made with Makie

5. Lastly, execute `x[]  = rand(100)` in a new cell to modify the values in the X-axis. The plot in the GLFW window gets updated thanks to the `Observables` mechanism supported by the recipe.

This section taught us how to create type and full plotting recipes with *Makie*. With this knowledge, you can start designing new plot types for the *Makie* ecosystem and add plotting capabilities to your Julia packages, if any.

Summary

This is the last chapter of this book on creating plots and interactive visualizations with Julia. Throughout this book, we have learned to make the most out of the `Plots` package and its backends. We have also explored other plotting packages from the Julia ecosystem, mainly *Makie* and `Gadfly`.

In the first part of this book, we learned how to use these packages, focusing on their interactivity. In particular, we explored the different topics using `Pluto` notebooks to benefit from their interactivity and reproducibility. In the last chapter of that section, we learned how to create engaging animations.

The second part of this book focused on exploring the Julia plotting ecosystem and different applications. First, we focused on data visualization and how to create classical statistical plots. Then, we explored graph analysis and visualization in Julia. After that, we learned how to visualize geographically distributed data. Finally, we learned about some tools in the ecosystem for plotting biological data, especially proteins and multiple sequence alignments. Almost all the visualizations that were explored in those chapters using the `Plots` package were based on the recipe system described in this chapter.

The last part of this book has dived deeper into the customization aspects offered by `Plots` – with hints regarding its *Makie* and `Gadfly` counterparts. We learned about the different parts of a plot and the attributes we can use to customize them. We also learned how to create plot themes to reuse those customizations and how to create complex layouts.

In this last chapter, we learned how to draw shapes, segments, and curves with `Plots` and *Makie*, while mainly focusing on the former. We also learned how to create custom plotting types using the recipe system of both plotting ecosystems. As a developer, you will find that last ability key to allow your users to plot your data types. Also, drawing shapes and defining recipes will allow you to create novel visualizations. For example, you can create a Donut chart by making a series recipe and drawing curved shapes.

Now that you've read this book, you can make the most out of Julia's plotting ecosystem to create the plots that better suit your needs.

Further reading

Creating plotting recipes with `Plots` is a critical topic that is covered in more depth in the following sources:

- `Plots` documentation: `https://docs.juliaplots.org/latest/recipes/`
- Daniel Schwabeneder's blog post: `https://daschw.github.io/recipes/`

You can read more on the *Makie* recipes system at `https://makie.juliaplots.org/stable/`.

You can find out more about Bézier curves by reading cormullion's blog post at `https://cormullion.github.io/pages/2018-06-20-bezier/`.

Index

Packt.com

Subscribe to our online digital library for full access to over 7,000 books and videos, as well as industry leading tools to help you plan your personal development and advance your career. For more information, please visit our website.

Why subscribe?

- Spend less time learning and more time coding with practical eBooks and Videos from over 4,000 industry professionals

- Improve your learning with Skill Plans built especially for you

- Get a free eBook or video every month

- Fully searchable for easy access to vital information

- Copy and paste, print, and bookmark content

Did you know that Packt offers eBook versions of every book published, with PDF and ePub files available? You can upgrade to the eBook version at packt.com and as a print book customer, you are entitled to a discount on the eBook copy. Get in touch with us at customercare@packtpub.com for more details.

At www.packt.com, you can also read a collection of free technical articles, sign up for a range of free newsletters, and receive exclusive discounts and offers on Packt books and eBooks.

Other Books You May Enjoy

If you enjoyed this book, you may be interested in these other books by Packt:

Learn Power BI - Second Edition

Greg Deckler

ISBN: 9781801811958

- Get up and running quickly with Power BI

- Understand and plan your business intelligence projects

- Connect to and transform data using Power Query

- Create data models optimized for analysis and reporting

- Perform simple and complex DAX calculations to enhance analysis

- Discover business insights and create professional reports

- Collaborate via Power BI dashboards, apps, goals, and scorecards

- Deploy and govern Power BI, including using deployment pipelines

Actionable Insights with Amazon QuickSight

Manos Samatas

ISBN: 9781801079297

- Understand the wider AWS analytics ecosystem and how QuickSight fits within it
- Set up and configure data sources with Amazon QuickSight
- Include custom controls and add interactivity to your BI application using parameters
- Add ML insights such as forecasting, anomaly detection, and narratives
- Explore patterns to automate operations using QuickSight APIs
- Create interactive dashboards and storytelling with Amazon QuickSight
- Design an embedded multi-tenant analytics architecture
- Focus on data permissions and how to manage Amazon QuickSight operations

Packt is searching for authors like you

If you're interested in becoming an author for Packt, please visit `authors.packtpub.com` and apply today. We have worked with thousands of developers and tech professionals, just like you, to help them share their insight with the global tech community. You can make a general application, apply for a specific hot topic that we are recruiting an author for, or submit your own idea.

Share Your Thoughts

Now you've finished *Interactive Visualization and Plotting with Julia*, we'd love to hear your thoughts! Scan the QR code below to go straight to the Amazon review page for this book and share your feedback or leave a review on the site that you purchased it from.

https://packt.link/r/1-801-81051-6

Your review is important to us and the tech community and will help us make sure we're delivering excellent quality content.

www.ingramcontent.com/pod-product-compliance
Lightning Source LLC
Chambersburg PA
CBHW060922060326
40690CB00041B/2972